Post-Tsunami Recovery in Thailand

Of all the huge natural disasters that claimed the lives of thousands in Asia, the Indian Ocean tsunami in 2004 was the largest, estimated to have killed more than 230,000 people. The scope of damage brought about by this natural disaster urges focus on recovery and post-disaster reconstruction from several perspectives. Here we find an in-depth ethnography of Thailand and the role of culture and religion as an underpinning issue in post-disaster recovery.

Following the post-tsunami recovery over five years, the book provides knowledge on socio-cultural responses from affected local communities after natural hazards, and is based on original material collected in Thailand after the 2004 tsunami. With a focus on how culture and religion interplay in the processes of building resilience and decreasing vulnerability, it gives a deeper understanding of how disasters are experienced and dealt with on a local level. It examines survivors' experiences of rituals and ceremonies that became a part of the survivors' lives in new ways after the tsunami, offering psychological reassurance and religious efficaciousness as well as communication links between themselves and the deceased.

Using observations, narratives and material from in-depth interviews with survivors, relatives, relief workers, officials and Buddhist monks and nuns, this book contributes to the research on anthropology of disaster and to the development of research on cultural resilience and religion in post-disaster recovery. It will be of interest to scholars of Disaster Studies, Buddhist Studies and Asian Studies.

Monica Lindberg Falk is a Social Anthropologist at Lund University, Sweden. She is the author of *Making Fields of Merit: Buddhist Female Ascetics and Gendered orders in Thailand* (2007), and has published several articles on themes related to gender, Buddhism, socially engaged Buddhism, and Buddhism and disasters.

The Modern Anthropology of Southeast Asia
Editors:
Victor T. King
University of Leeds
Michael Hitchcock
University of Chichester

The books in this series address issues in processes of development, globalisation and change in Southeast Asia. Where appropriate, they contextualise change and local responses to it by providing ethnographic materials on social and cultural forms and institutions. Although all the contributors to the series examine modern and contemporary issues in the anthropology of Southeast Asia, the emphasis in each book differs as authors choose to concentrate on specific dimensions of change and globalisation or work out particular conceptual approaches to the complex issues of development. Areas of concern include: nation-building, power and the media; technological innovations in agriculture and rural-urban migration; the expansion of industrial and commercial employment; the rapid increase in cultural and ethnic tourism; the consequences of deforestation and environmental degradation; heritage and identity; contemporary expressions of religious afflliliation; the 'modernisation of tradition'; ethnic identity and conflict; changing gender relations; and the religious transformation of society.

The Modern Anthropology of Southeast Asia: An Introduction
Victor T. King and William D. Wilder

The Changing Village Environment in the Southeast
Applied anthropology and environment reclamation in the northern Philippines
Ben J. Wallace

The Changing World of Bali
Religion, society and tourism
Leo Howe

Modernity and Malaysia
Settling the Menraq Forest nomads
Alberto Gomes

The Orang Suku Laut of Riau, Indonesia
The inalienable gift of territory
Cynthia Chou

Southeast Asian Perspectives on Power
Liana Chua, Joanna Cook, Nicholas Long and Lee Wilson

Sex, Love and Money in Cambodia
Professional girlfriends and transactional relationships
Heidi Hoefinger

Identity and the State in Malaysia
Fausto Barlocco

Democracy, Corruption and the Politics of Spirits in Contemporary Indonesia
Nils Bubandt

Post-Tsunami Recovery in Thailand
Socio-cultural responses
Monica Lindberg Falk

Post-Tsunami Recovery in Thailand

Socio-cultural responses

Monica Lindberg Falk

Routledge
Taylor & Francis Group
LONDON AND NEW YORK

First published 2015
by Routledge
2 Park Square, Milton Park, Abingdon, Oxon OX14 4RN

and by Routledge
711 Third Avenue, New York, NY 10017

Routledge is an imprint of the Taylor & Francis Group, an informa business

British Library Cataloguing in Publication Data
A catalogue record for this book is available from the British Library

Library of Congress Cataloging in Publication Data
Falk, Monica Lindberg, author.
Post-tsunami recovery in Thailand : socio-cultural responses / Monica Lindberg Falk.
pages cm. – (The modern anthropology of Southeast Asia)
Includes bibliographical references and index.
1. Tsunami relief–Social aspects–Thailand. 2. Disaster victims–Rehabilitation–Thailand. 3. Indian Ocean Tsunami, 2004–Social aspects. I. Title.
HV6032004.T5 F35 2014
363.34'94809593–dc23 2014011589

ISBN: 978-1-138-02357-4 (hbk)
ISBN: 978-1-315-77632-3 (ebk)

Typeset in Times New Roman
by Wearset Ltd, Boldon, Tyne and Wear

Printed and bound in Great Britain by
TJ International Ltd, Padstow, Cornwall

This book is dedicated to the survivors and the memory of those who lost their lives in the Indian Ocean Tsunami of 26 December 2004.

Contents

Acknowledgements

This book is a product of several years of research and I am indebted to a large number of people without whom this work would not have been possible. My first thank you goes to the survivors of the disaster, their relatives, the Buddhist monks, nuns, villagers and aid workers for generously sharing their knowledge, experiences and everyday life with me. I would like to thank the *mae chi*'s relatives in Phuket with whom the *mae chi*s and I stayed when we visited Phuket and Phang Nga in January 2005. I would also like to thank the family who invited my assistant and me to stay in their "knock-down" temporary shelter when they had moved to a house in the village, and also for their help to rent a partly damaged house in Nam Khem. I particularly want to thank my assistant Nantana Pidtong. Without her knowledge, patience, integrity, empathy and skills in working in a post-disaster situation, this study would have been difficult to carry out.

I wish to thank my colleagues at the Centre for East and South-East Asian Studies, Lund University, for taking an interest in my work and their appreciated comments. I also wish to thank Mahidol University International College for hosting the project and I am grateful to my Thai colleagues for fruitful discussions and insightful comments on findings from this study. I owe special thanks to Hiroko Kawanami, Samuel Geoffrey, Anne Guillou, Silvia Vignato, Jean-Christophe Gaillard, Pauline Texier, Leslie E. Sponsel, Marja-Liisa Keinänen, Sunita Reddy and Harish Naraindas for organizing panels at conferences and for valuable comments on my papers based on the material from this post-tsunami research. The manuscript has benefited from close readings and constructive comments by Routledge anonymous reviewers and I extend my thanks to the editorial team at Routledge. I also want to thank Kay Hutchings for her patience and for proofreading the text.

The research project was made possible by generous financial support from the Swedish Research Council (VR).

My gratitude also extends to many friends in Thailand who offered friendship, advice and practical assistance. I would also like to express my appreciation to friends and colleagues in Sweden who have been interested in this study.

Some material presented in this book has previously been published. In Chapter 2, passages are reproduced from "Gender, Buddhism and Social

Resilience in the Aftermath of the Tsunami in Thailand" published in *South East Asia Research*, 2 (2012) 175–190, copyright 2012 SOAS, reproduced by permission of IP Publishing Ltd. In Chapter 5, passages are reproduced from "Recovery and Buddhist Practices in the Aftermath of the Tsunami in southern Thailand" published in *Religion*, 40 (2010) 96–103 and permission to reprint has been granted.

Finally, this project could not have been completed without the love and support of my family and close friends. I am deeply grateful to my husband Bengt-Arne, my daughter Maria, her husband Henrik and their daughters, who have patiently put up with my research travels, conference trips and my absorption in this project.

Abbreviations

ADRA	Adventist Development and Relief Agency
APWLD	The Asia Pacific Forum on Women, Law and Development
DOMH	The Mental Health Department
DVI	Disaster Victim Identification
INGO	International non-governmental organization
IRC	International Repatriation Centre
MOPH	The Ministry of Public Health
NGO	Non-governmental organizations
NIDA	National Institute Development Administration
PTSD	Post-Traumatic Stress Disorder
TAT	Tourism Authority of Thailand
TTVI	Thai Tsunami Victim Identification
UNDP	United Nations Development Programme
UNESCO	United Nations Educational, Scientific and Cultural Organization
UNICEF	United Nations Children's Fund

1 Introduction

The 2004 Indian Ocean tsunami

In recent years a number of natural hazards has struck the Asian region and they have been devastating for the communities and the people affected. Of all the huge natural disasters that claimed the lives of thousands, the Indian Ocean tsunami in 2004 was the largest, estimated to have killed more than 230,000 people. In 2005 more than 70,000 lost their lives following the earthquake in the Kashmir region of India and Pakistan. In 2008 about 130,000 people perished in Cyclone Nargis in Burma/Myanmar, and in the same year more than 80,000 people died following the earthquake in Sichuan in China. In 2011 the so-called triple disaster in Japan struck, with an earthquake followed by a tsunami, which triggered the nuclear power plant disaster that caused about 20,000 people's deaths and in 2013, Typhoon Haiyan killed about 7,800 people in the Philippines. In the Asian region millions of people's lives are affected yearly by natural hazards such as flooding, typhoons and droughts. It is well known that disaster risks are higher in poor countries and that it is usually poorer communities that are especially hard-hit when disasters occur (ISDR 2009). Following a disaster, most of the published reports, surveys, evaluations and articles focus on the material aid assistance carried out after the catastrophe and most of them are prepared by international aid workers with limited knowledge and experience of the local area. A natural hazard is, first and foremost, a local matter and how people cope with disasters depends largely on the socio-cultural environment (Oliver-Smith and Hoffman 2002). The scope of damage brought about by disastrous events in Asia urges focus on recovery and post-disaster reconstruction from several perspectives.

The 2004 Indian Ocean tsunami was a natural disaster of global proportions. The magnitude of the destruction ranks as one of the greatest natural disasters in recent history. Following the tsunami, there was a tremendous response from individuals and aid organizations across the world (see e.g. Clarke *et al.* 2010). Large aid programmes were proclaimed in support of relief and reconstruction. The aid and relief operation was a complex task and one of the biggest recovery and reconstruction efforts that had been carried out (Jayasuriya *et al.* 2010). The focus of this monograph is on culture and religion in the recovery process and addresses culture as an underpinning issue in post-disaster recovery. It concentrates on socio-cultural responses from affected local communities after natural hazards. The study explores Buddhist and spiritual beliefs in the recovery process and gives examples of how Thai Buddhist monks, nuns and laypeople employ Buddhism in order to cope after losing family members, friends, neighbours, their homes and means of livelihood. The process of reconstruction involves both restoration and change and the ethnography in this monograph is based on material collected in Thailand after the 2004 Indian Ocean tsunami. Theoretically, the book aims to give a deeper understanding of how disasters are experienced and dealt with on a local level, with a focus on how culture and religion interplay in the processes of building resilience and decreasing vulnerability. The monograph explores local adaptation strategies, investigates human resilience as a subjective process of recovery and gives nuanced aspects of experiences of disaster and recovery. It contains observations, narratives and material from in-depth interviews with survivors, relatives, relief workers, officials and Buddhist monks and nuns.

Methodological approach

This anthropological study followed the post-tsunami recovery over five years. It was carried out in small fishing communities affected by the tsunami located mainly in Phang Nga, the worst-hit province in Thailand.[1] I had experiences of conducting ethnographic fieldwork in central, north and northeast regions of Thailand and this was the first time I carried out fieldwork in the south of Thailand. A few weeks after the tsunami struck, I visited the disaster-stricken areas together with a small group of Thai Buddhist nuns, *mae chi*s, from central Thailand that I have known for many years. In the following years, I tracked the survivors' recovery process in several villages situated mostly in Phang Nga province. I employed an ethnographic, anthropological holistic approach in order to understand in what ways Buddhism was involved and important in the recovery process. I gathered narratives about people's experiences of the tsunami and their recovery. Temples and temporary shelters were initially obvious sites for collecting information and I met many of my informants there. Through informants' networks I came into contact with survivors whom I would otherwise not have encountered, and some of them lived in remote places. Most of the informants were between 20 and 50 years old, but I also interviewed older people. I interviewed both men and women, and almost all of the lay informants had lost people close to them in the disaster. Most of the informants' houses were totally or partially damaged by the tsunami and the majority had lost their livelihoods. I interviewed monks and nuns who were permanent residents at the temples in the affected areas and those who had travelled to the afflicted areas to help the local temples after the disaster.

During the course of the fieldwork my main assistant and I were involved in the daily life of the villages and were invited to various social events and ceremonies. We met with people in their homes and in their villages. Laypeople invited us to events and ceremonies that were held. We were in close contact with several Buddhist temples in the affected areas and we regularly attended ceremonies and activities that took place at the temples and participated in the remembrance ceremonies after the tsunami. We frequently visited self-help groups and non-governmental organizations (NGOs) working on the recovery process after the tsunami.

Due to the complexity of the research site and the sensitive nature of working with traumatized informants, a number of different research techniques were utilized. In 2005, I began this project by making a general survey of temples in the affected area, which gave me an overview of the post-disaster situation. I decided to focus on temples that were actively involved in assisting the survivors in the post-tsunami situation and chose temples and survivors in five villages. Informants were selected both among those who frequently visited temples and those who, for various reasons, did not often go to them. I talked with laypeople at the temporary shelters and in the villages on islands, as well as on the mainland. The initial phase of collecting basic information from the temples and temporary shelters provided some opportunities for in-depth interviews, but a further strategy for obtaining detailed interviews was through social networks. The major techniques involved were participant observation, semi-structured in-depth interviews, life stories and narratives. The research also included informal talks with survivors, Buddhist monks and nuns, and family members and relatives of survivors. Additional data was supplemented with interviews with mental health doctors and relief workers who had valuable information about the recovery process.

Perspectives on death, loss and the roles of religion in recovery after a catastrophe are highlighted in this study. The significance of Buddhist monks and nuns in consoling people, explaining life and death, performing ceremonies and restoring continuity and order is explored. The book analyses collective ceremonies and self-help groups, which have proven to be important in decreasing feelings of loneliness, helplessness and pain. It also examines survivors' experiences of rituals and ceremonies that became a part of their lives in new ways after the tsunami, offering psychological reassurance and religious efficaciousness as well as communication links between themselves and the deceased.

Since the study is primarily based on information collected from personal observation and interviews, I have followed ethical standards and masked the informants' identities, therefore personal names in this book are fictitious. Working with Thai assistants was invaluable. During the five years of research, I spent periods of one to several months at a time in the field. After the tsunami the region was swamped with foreigners who, in various capacities, had come to help out in the time of crisis, and at the start of the fieldwork I was often mistaken for an aid worker. It was through working together with my main assistant who knew the area, staying in the field for a long time and getting to know the informants, that we could build trust and get more in-depth information.

The local setting

Phang Nga province has been the main site for fieldwork and the population in the coastal areas is multi-ethnic (see Figure 1.1). Many of those who were

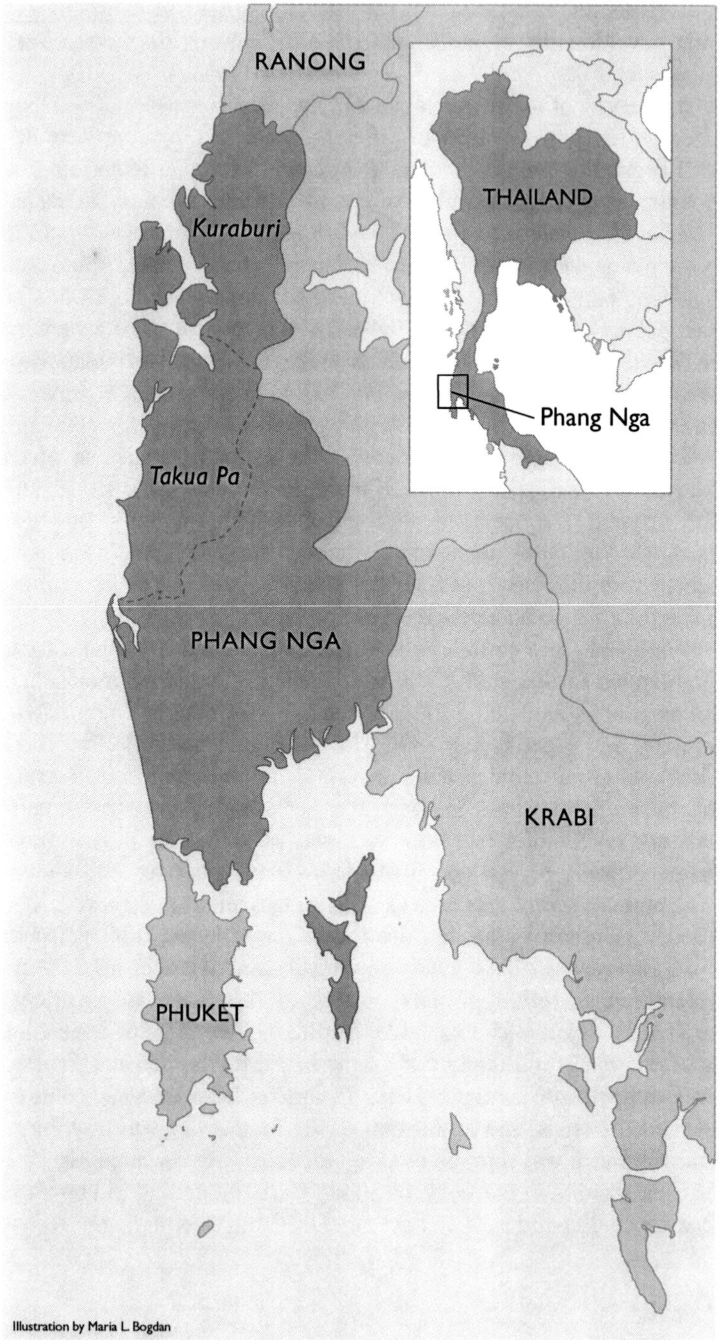

Figure 1.1 Map of Phang Nga (source: illustration by Maria L. Bogdan).

affected by the tsunami were fishing families and migrant workers who lived in impoverished circumstances. From a regional perspective, poverty is not predominant in the south of Thailand. However, the provincial variations in living standards are starker than the regional differences and the disparities within provinces on sub-district levels are striking. Provinces with low poverty rates, such as Phuket, exist alongside provinces with higher poverty rates like Ranong and Phang Nga. Phang Nga has quite recently developed into a growing tourist destination and the livelihoods of people in the villages close to the Andaman Sea have become increasingly tourism-based (see Calgaro *et al.* 2009). The expanding tourist industry in Phang Nga attracts migrant workers on a large scale from other parts of Thailand and also labourers from neighbouring countries, particularly Myanmar.

Phang Nga province

In Thailand, Phang Nga was the worst-hit province by the tsunami. A brief outline of the area is provided to afford a better understanding of the area's circumstances. Phang Nga is located about 800 km from Bangkok on the coast of the Andaman Sea. It has mountainous forests on the mainland and the coastal line in the west is 240 km in length; over 105 islands belong to the province. Phang Nga is divided into eight districts and 48 sub-districts containing 321 villages. Agriculture is a main source of income and economic crops in the area are, for example, para rubber, oil palm and rambutan. Fishing has been the economic mainstay of the majority of the local population along the peninsular coast and on the many islands. Prior to the growing tourism industry, the coastal areas were sparsely settled by fishing villages, settlements of former mining families and settlements of the ethnic, semi-nomadic groups Moken and Moklen. Tourism has attracted a growing number of people from other regions of the country as operators and employees of enterprises and services, as well as legal and illegal Burmese/Myanmar migrant labourers (Cohen 2008: 28).

In February 2007, Phang Nga's population was 246,094. In one district, all the villagers were Buddhists and in another district all the villagers were Muslims. In the other six districts, the villagers' religious affiliation was mixed: 10–30 per cent were Muslims and 70–90 per cent were Buddhists.[2] The Moken and Moklen people have lived in the area for more than a century and have a distinctive culture of ancestor worship, although some have simultaneously had long and close connections to the Buddhist temples. After the tsunami, groups of Moken and Moklen were targeted by Christian missionaries and some of the groups converted to Christianity. The French anthropologist Jacques Ivanoff said that outside pressures such as those from faith-based aid groups could "threaten their way of life more than any number of tsunamis".[3]

As well as agriculture and fishing, the south of Thailand has a long history of tin mining. In the beginning of the nineteenth century huge amounts of tin deposits were found in many districts in Phang Nga, among them Kura Buri and Takua Pa districts. At an early stage in the development of the tin mining industry, families

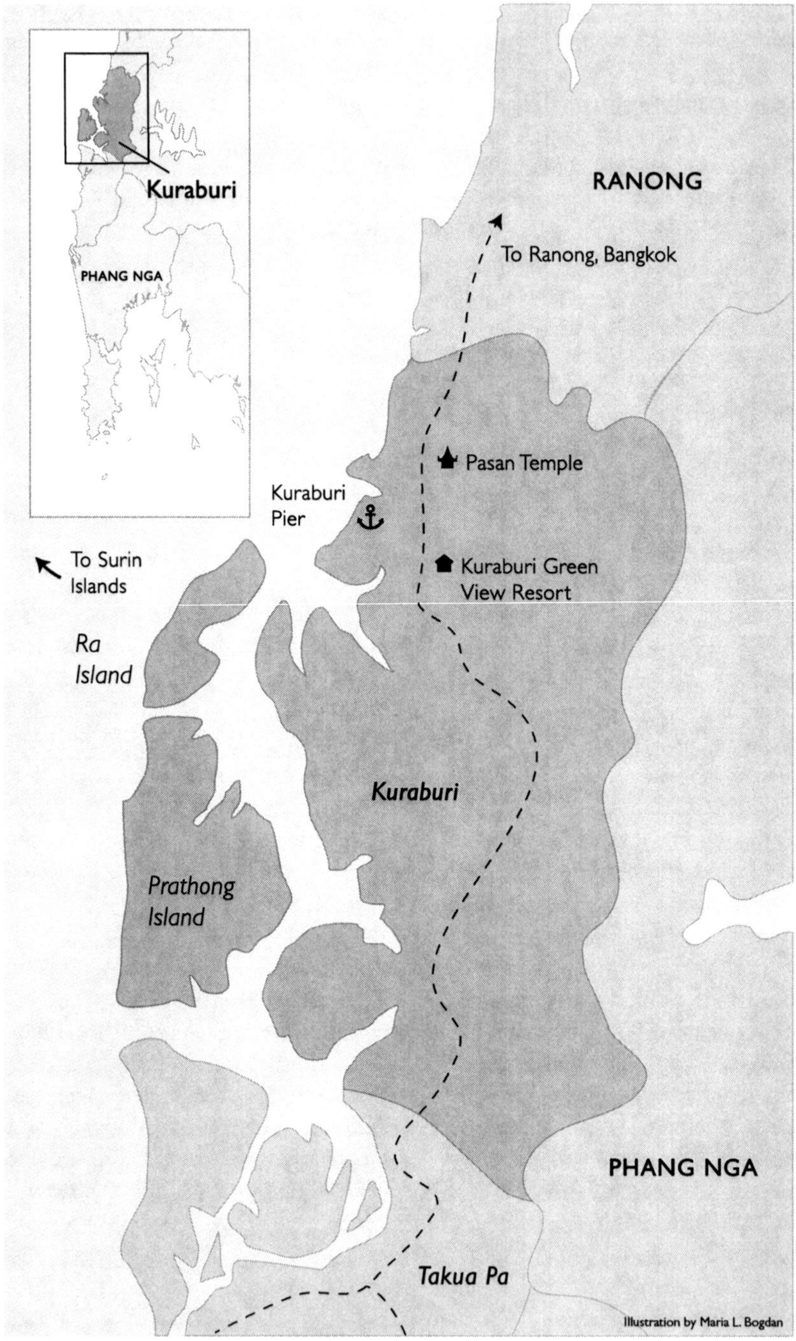

Figure 1.2 Map of Kura Buri (source: illustration by Maria L. Bogdan).

of Chinese origin based in Penang (in Malaya), gained concessions to mine and export the tin. These families also owned smelters and steamers to ship the tin. Chinese coolies who worked with tin in Malaysia (Malaya) also worked in the tin mines in Thailand (Siam) (Lim Kwee Phaik 2011). The coastal environment became affected by the tin mining, and reefs in the bays were destroyed. Moreover, mangrove concessions had been granted to loggers from the charcoal industry, and that reduced the shore's natural protection against devastation (Warunpitikul and Tangwisutijit 2005).

Takua Pa district and town

Takua Pa district has a history as an important centre of tin mining and the region increased in importance after the Dutch, British and Portuguese arrived to prospect for tin ore. Along with the European prospectors came Chinese labour, to work the mines in Phang Nga and neighbouring Phuket province. Tax records from that time show that the tin mining business hit a peak at the end of 1870s (Lim Kwee Phaik 2011). When tin prices dropped, the industry went into a gradual decline, followed by an exodus of Western investors who sold off or left their businesses in the hands of their Chinese employees. These workers later married into local families and their descendants constitute a sizeable part of Takua Pa's and Phuket's populations today. The end of the tin industry in south Thailand came in 1985 when all the mines were shut down. Unfortunately, these mines, the very source of wealth for several families, have since become an environmental liability (ibid.). Several artificial mounds and ponds formed by previous tin mining activity are to be found, for example, in Ban Nam Khem, the village that was hit worst by the tsunami.

Takua Pa town is the administrative capital of Takua Pa district. The old part of the town is located in the south of the town, near Takua Pa's new town, Yan Yao, became well known after the tsunami for the key role played by the Wat Yan Yao temple as a centre for the identification of corpses. The coastal area in Takua Pa district was a newly developed tourist centre when the disaster struck. Many hotels and cottages had been built in few years before the tsunami happened. Khao Lak in Takua Pa district was one of the worst-affected tourist destinations (Calgaro *et al.* 2009: ix). The relatively new tourist location, Khao Lak in Phang Nga, had in 2000 established charter tourism, with visitors mainly coming from Northern Europe. As a tourist destination, Khao Lak was made up of several small communities located on about 15 km of flat land, with beaches and hotels extending from the seashore to a couple of kilometres inland. The islands Koh Phra Thong and Surin are situated north of Khao Lak in Kura Buri district and had not developed charter tourism in 2000. In that area there were more so-called eco-tourists and individual tourists who stayed in guesthouses and "home stay" accommodation, provided by local households.

In 1988 the first ten bungalows for tourists were built in Khao Lak. The area remained an undiscovered tourist destination until 1996 when Khao Lak Laguna Resort reached out to the market and tourists became aware of the serene location. The tourist agencies named the area Khao Lak after a mountain close to the tourist region. The area attracted both Thai and foreign investors and for many of the local families living in the area, tourism had developed into their main source of livelihood. Khao Lak grew rapidly from having only a total of 100 rooms for tourists in 1996 to 5,312 rooms in December 2004 (Calgaro *et al.* 2009: 15). When the tsunami hit, Khao Lak had a mix of larger resorts, hotels, small bungalows, restaurants and shops. The tourist business became a magnet for workers from the greater area in Phang Nga and also from other places in Thailand as well as for migrant workers from neighbouring countries.

Ban Nam Khem village

Ban Nam Khem, the most seriously damaged village in Thailand, is located approximately 120 km north of Phuket and about 10 km west of Takua Pa town. Ban Nam Khem was described by the villagers as a small easy-going fishing village with a simple way of life. The village had grown from being a small community consisting of only four or five families to a thriving village with a population of over 5,000 people. The village's growth was the result of the emerging tin mining industry, fishing and, more recently, tourism (Ivanoff 2005).

Back in 1927, a few families made a living in Ban Nam Khem by working as fishermen. Some also grew vegetables and raised animals. In 1933 some Moken families settled in Ban Nam Khem, and around that time Japanese soldiers distributed goods to villagers and built piers in a section now known as Leam Pom (ibid.). A canal was also dug to allow small fishing boats access inland. In 1956 rich tin deposits were discovered in Ban Nam Khem, bringing about the start of the village's tin mining period. Labourers worked with spades to dig out the tin and companies came to buy the extracted tin from them. Eventually, a company employed labourers from other provinces to work in Ban Nam Khem's tin mines.[4] During that time, electricity was brought to the village and the villagers built a school and a Buddhist temple.

By the 1970s, Ban Nam Khem was a prosperous community, playing an important role in Phang Nga's tin mining industry. The industry's demand for labour attracted a broader population, but also introduced an increase in criminal and illicit behaviour. Informants mentioned how dangerous Ban Nam Khem could be and it was common to find murder victims on the beach in the morning. Today, the traces of the mining period include some tin mining mounds (with a maximum height of over 10 m) distributed over the plains, and several ponds derived from tin mining are also located in the lowlands.

The beginning of the region's tourism industry was also associated with the village's growing economy. In the 1980s and 1990s, as the world's tin market declined, the local economy began to rely primarily on the established fishing and tourism industries. With a steady population of nearly 5,000 residents, Ban Nam Khem remained a flourishing village until the tsunami in December 2004.

Ethnic minorities at Koh Phra Thong

The island of Koh Phra Thong, which belongs to Kura Buri district located north of Takua Pa, was badly hit by the tsunami. South of Koh Phra Thong is the island Koh Kho Khao, and both are flat open islands with long sandy beaches. North of Koh Phra Thong is the island Koh Ra, a mountainous island with rain-forest. Koh Phra Thong is multi-ethnic with Moken, Moklen, Burmese, Thais and Sino-Thai inhabitants.

The Moken live along the seashore and on the islands off the western coast of peninsular Myanmar (Elphinstone Island) and off the Mergui Archipelago off Surin Island in Thailand. The Moken are characterized as semi-sea nomads since some of them today have settlements, for example, on Surin Island and their population is estimated to be about 3,000. The Moklen group also consists of about 3,000 individuals and they live in small communities along the coast of the provinces Phang Nga and Phuket. They are also known by the name "new Thais" (*mai thai*).

Most Moklen groups are concentrated along the coast between Takua Pa and Thai Muang in Phang Nga province. There are over 20 Moklen villages between Koh Phra Thong and Phuket (Ferrari *et al.* 2006). The Moken and Moklen began to live on Koh Phra Thong in the 1930s. They had known the environment for generations when some of them decided to live permanently on the island (ibid.: 23–25). The Moken and Moklen are known for their diving and navigating expertise and skills (Weber 2005: 11). Jacques Ivanoff has observed that the Moken are not fishermen because they use no fishing techniques. Instead of eating fish, their main food consists of molluscs, oysters and sea urchins and their diet depends on the tidal rhythms (Ivanoff 1997: 43). The decision to reside more permanently on the island has implied a change in their livelihood and they commonly have seasonal employment organized by local middlemen, *taukay* (traditionally Chinese traders). They work with shrimp production, small-scale fisheries, road mending, rubber cultivation, farming and the like, and they have also worked in the tin industry. The Moklen use the Thai term *luk chang*, "labourers", to describe their means of employment.

The Moken are dependent on the sea. Their traditional way of life was sea-based, whereby between ten and 30 boats with families would join together and sail from island to island. They brought up their children on their boats, ate and slept on the boats and only left the vessels when they went to search for food. Even when they took refuge during the monsoon in a protected bay, and some families built small huts of bamboo and leaves, many of them still lived on the boats, which were moored on the shore (Ivanoff 2005: 16).

In 1968, the royal Princess, Somdet Ya (the royal grandmother) visited Koh Kho Khao and Koh Phra Thong and that resulted in all of the Moklen receiving "identity papers"; they were also given the surnames that they carry today. Three thousand Moklen share the following four family names: *Klatalay* which means "audacious fearless men of the sea", *Hantalay*, which means "courageous of the sea", *Nawarak*, which means "water" and *Samutwari*, which means "ocean, water". The Moklen's identity was thereby recognized, as was their right of residence. The Moken who were on the islands at the time of the royal visit were given the name *Klatalay* and they also obtained identity papers (Ferrari 2006: 18). By giving the Moklen family names, the royal princess authorized them to possess identity cards and opened the doors for them to become Thai citizens; they also gained the right to education, health care and, in a certain way, access to property. A monk in Kura Buri district told me about the occasion when Somdet Ya gave the Moklen their surnames. He said that it happened almost 50 years ago when the princess visited Phang Nga. The monk told me that the princess was at sea in a boat, when the Moklen surrounded her with their vessels. She asked about their names and when she discovered that they did not have any surnames she gave them last names that were easy to distinguish from other Thai surnames.[5]

In the mid 1980s a group of 200 Moken people lived on the island and in National Park Koh Surin (whose administration depends on the district of Koh Phra Thong). The officials had tried to expel the Moken from the territory without success. In 1986, Somdet Ya, who had shown an earlier interest in the ethnic minority groups' situation, came to officially open the national park and to include the park as a part of the nation. She was able to intervene so that the Moken group could continue living on Surin Island and she ensured that they could have ten boats at their disposal (Ferrari 2006: 17).

The impact of the tsunami

The Indian Ocean tsunami affected 13 countries and the nations most affected were Indonesia, India, Sri Lanka and Thailand. About 230,000 people were killed and more than one million people were displaced. Vast areas were totally destroyed; many of those who survived lost their homes, families, friends and neighbours, and a large number of children lost their parents. The tsunami was a unique catastrophe in many ways because it affected Thai people from several ethnic groups, illegal immigrants from neighbouring countries, poor people from other parts of Thailand, affluent tourists from countries abroad and people of all ages. Many of those who were directly affected were poor fishing families and migrant workers.

Injuries and loss of life in the tsunami in Thailand

The tsunami hit the six Thai provinces: Ranong, Phang Nga, Phuket, Krabi, Trang and Satun. The waves were more than 12 m high and many of the provinces' coastal areas and islands were badly affected (Umitsu 2011: 39). In Thailand, the tsunami affected 25 districts and 95 sub-districts with a total population of 1.9 million people. The waves damaged 412 villages, of which 47 were almost completely destroyed,[6] and caused losses, damages and suffering to an extent that could hardly be understood. Efforts were made to measure the disastrous impact of the tsunami. Figures detailing the death toll and destruction mentioned in media and other reports varied widely and the estimations most likely understated the scale of the catastrophe. In interviews that I conducted with local people, they often mentioned that the numbers of deaths presented by authorities seemed to them to be too low.

An estimate carried out about one year after the tsunami stated that the number of people who lost a family member or home was about 100,000 (UN 2005: 13). The number of corpses found along the coast and near the islands was 5,395. Among the identified dead were 2,059 Thai and 2,436 foreigners; 900 corpses were unidentified. The number of people registered as missing was 2,817: 1,921 were Thai and 896 were foreigners. Assuming that they perished, the number of dead people amounted to 8,212 (see Table 1.1). The number of Thai victims then rose to 3,980 and that of deceased foreigners to 3,332. According to the records, about 8,457 people had suffered injuries that required medical treatment and they included 6,065 Thai and 2,392 foreigners (ibid.: 14).

About 20 per cent of schools, serving 50,000 children in 20 districts, were affected by the tsunami (UNICEF 2005). Five schools were destroyed and 19 more were damaged; many schools lost furniture and equipment. Quite remarkably, however, all the schools had reopened by 10 January 2005, with the majority having already begun their school year on 4 January, and 75 per cent of the children were back at school during the first week.

In the six provinces hit by the tsunami, 1,449 children had become orphans. They comprised children who lost one or both parents, or their immediate guardian and supporter.[7] The initial help to each orphan who had confirmed evidence such as the death certificate of parents or supporters was 1,000 baht per person (about US$30). If the evidence was not clear, officers would visit the house and conduct interviews to ascertain the validity of the request for aid. Associations, offices, foreign organizations and individuals helped the orphans by donating money or educational support (Weber 2005: 62).

Ethnic groups and migrant workers

More than 400 fishing communities were affected by the disaster. Many of those who were already poor suffered particularly from the devastation that the

Table 1.1 Deaths, injured and missing after the tsunami in Thailand

Province	*Dead*	*Injured*	*Missing*
Krabi	721	1,376	544
Phang Nga	4,225	5,597	1,655
Phuket	279	1,111	608
Ranong	159	246	9
Satun	6	15	–
Trang	5	112	–
Total	5,395	8,457	2,817

Source: UN (2005: 14).

tsunami brought. Fishing boats were destroyed or damaged, fishing gear was lost and all hatcheries along the coast were washed away. Among the hardest hit were the Moken, Moklen and Urak Lawoi ethnic groups. About 60 per cent of these ethnic minority groups' households were affected (UN 2005: 15). These groups and fishing communities had been living for decades on "public" land and thus did not have not secure tenure and access to basic services. The complex problems of their land rights and resettlement made their rehabilitation and recovery much harder (ibid.: 16).

The effects of the tsunami were magnified for those most vulnerable. One of these groups was migrant workers, the majority of whom were from Myanmar. Rough estimates show that at least 7,000 migrant workers and their dependents were affected by the disaster. This estimate is based on the number of registered migrants who were employed in sectors that were most damaged: fishing, construction and tourism. However, given that for every registered migrant there were an estimated two or more unregistered migrants, the real number of those impacted by the tsunami might indeed be much higher. In the aftermath of the disaster, migrant workers and their families became particularly vulnerable because of their uncertain immigration status. Crucial identification and employment documents were lost in the waves and left the migrants in a state of legal uncertainty. Instances of arrest and deportation created concern within the migrant community, causing many to hide or flee inland and further depriving them of much-needed humanitarian and medical aid. For some, the fear was so great that they did not go to official morgues to identify their relatives (UN 2005: 16).

The impact of the tsunami in Phang Nga

In Phang Nga province, six districts were struck by the tsunami: Kura Buri, Takua Pa, Thai Muang, Takua Thung, Koa Yao and Muang Phang Nga districts. The affected area covered 19 sub-districts and 69 villages. In Thailand the death toll was highest in Phang Nga province and the official number of dead in the province was 4,225, (1,389 Thai, 2,114 foreigners, 722 unidentified). In September 2005, 1,696 people were still missing (Phang Nga: 1st Year Tsunami Memorial 2005: 5). According to Phang Nga province's records, 5,597 people were injured.

Takua Pa was the hardest-hit district; the giant waves caused tremendous loss of life and the most significant damage in 17 villages located in three of the region's eight sub-districts. Ban Nam Khem, located in the northwestern part of the plain, was the most severely hit village, with the tsunami wave striking the village directly from the west and northwest. The death toll in Takua Pa district was exceedingly high and in the three coastal sub-districts, 3,808 people died. Among the survivors, 4,210 were registered as injured and received medical treatment. In Takua Pa 1,749 families and 8,917 people survived the tsunami but were severely impacted (Phang Nga: 1st Year Tsunami Memorial 2005: 5).

After the tsunami, the local hospitals initially thought that they could handle the situation without outside help. However, after a couple of hours the hospitals were overcrowded with wounded people and the hospitals' managers asked for assistance. The Ministry of Public Health (MOPH) mobilized more than 200 doctors, who immediately moved to the affected areas. The health system responded effectively, with the distribution of clean water and the immunization of children as well as the setting up of a disease surveillance system (UN 2005: 25). Some hospitals sent mobile medical teams to help out in the affected areas and some took turns providing medical help to the local hospitals. Others transferred patients from the hospitals in Phang Nga, Phuket, Krabi and Ranong to their own hospitals for further medical treatment. Not only did hospitals from nearby provinces provide help, but hospitals from central Thailand also assisted their colleagues in the south. The MOPH established a command centre in Phuket to coordinate health services, epidemic surveillance and emergency medical response for the impacted areas and it also set up health centres at hospitals in Krabi and Phang Nga provinces, with teams of physicians and volunteers from outside the area providing help. The hospitals organized volunteer register systems to make best use of volunteers' specializations. Most of the patients had wounds all over the body or on parts of the body. The top five types of injury were open wounds and laceration of soft tissues down to the muscles on the arms, legs, chest, abdomen and back; drowning; dying before reaching the hospital; and pneumonia, typically resulting from choking on water, mud and sand. Almost all of the victims suffered infections and multiple fractures of the arms, legs, ribs, wrists and ankles (Wanaratna 2005: 13). The Thai Government led an effective emergency response in terms of relief supplies and health service. No outbreaks of communicable diseases were reported and that was a result of the effective emergency action.

Material damage

The tsunami, with its giant waves, ruined many villages and several thousand houses in Phang Nga province. Phang Nga sustained the worst damage with regards to destroyed or damaged public utilities such as piers, bridges, roads, drains, dikes, dams, the electricity grid, water supply and the telecommunications network (Weber 2005: 35). Kura Buri district, followed by Takua Pa, Thai Muang, Takua Thung and Koa Yao districts suffered the greatest damages to agricultural land when the tsunami destroyed cropping areas which were valued highest in monetary terms (ibid.: 35). Throughout the coastal areas of Phang Nga province, 6,493 members of fishing communities reported material losses. Two major ports in Phang Nga province, Ban Thap Lamu in Lam Kaen sub-district in Thai Muang district and Ban Nam Khem in Bang Muang sub-district in Takua Pa district were severely damaged. The giant waves devastated buildings and destroyed fish ponds, fishing boats and equipment (Weber 2005: 36). Cattle, sheep, goats, pigs and poultry were lost. Rice fields and gardens, restaurants and shops were devastated. Ninety-three hotels, piers, bridges, drainage systems, electricity and water supplies and telephone systems were damaged. Moreover, mangrove forests, regular forests and coral reefs were destroyed (Phang Nga: 1st Year Tsunami Memorial 2005: 6).

The waves tossed the debris of buildings and cars across roads and beaches. All kinds of buildings were devastated by the tsunami, leaving thousands of people homeless. Likewise affected were schools, Buddhist temples, Muslim mosques and public spaces. The waves destroyed or damaged hotels and shops along approximately 50 beaches. A great many of the characteristic bungalows, rural restaurants, shops and entertainment spots built for tourists were ruined. Many buildings constructed using modern technology and materials did not withstand the forceful waves. Damage and destruction of the infrastructure made the rescue operations after the tsunami complicated. The mobile phone network was overloaded and shut down and Phuket International Airport cancelled all flights.

In Phang Nga, five schools in Ban Bang Sak, Takua Pa district; Ban Thap Lamu, Thai Muang district; and Bang Pak Chok, Kura Buri district were totally destroyed. The school buildings in Ban Nam Khem, Takua Pa district, and Ban Thung Dap, Kura Buri district, were severely damaged and 1,021 students were left without premises.

Along a 120 km strip of beaches in the districts of Kura Buri, Takua Pa, Thai Muang, Takua Thung and Ko Yao, almost all tourism resorts were ruined. The tourism industry in Phang Nga province sustained the worst impact of the provinces on the Andaman Sea coast. At Khao Lak in particular, all types of accommodation were devastated, including hotel buildings, bungalow-style resorts, hostel-type structures and guest-houses, as well as restaurants, boutiques, food-stalls and assorted shops (Calgaro *et al.* 2009).

Devastated livelihoods

The coastal area of Phang Nga province, with its marine resources, has for many years attracted people who wish to establish a livelihood. In addition to the largely self-employed, economically active members of the farming and fishing communities, provinces such as Phang Nga had a large employee labour force. Many locals and large numbers of migrant workers from other regions of Thailand, as well as from neighbouring countries, were employed in either the tourism or construction industries.

The tsunami not only swept away the jobs of 120,000 workers in the fishing communities, it also made the quality of life of the more vulnerable ethnic groups and migrant workers even more perilous. The Moken, Moklen, old Thai communities engaged in coastal fishery, staff of aquaculture enterprises and crew members of commercial fishing trawlers were among the survivors (Phang Nga: 1st Year Tsunami Memorial 2005: 23).

One of the worst-hit areas of Khao Lak was Laem Pakarang, which was hit with waves of more than 11 m in height, pushing more than 2 km inland. Calgaro *et al.* (2009), concluded that Khao Lak's heavy dependence on tourism is due to few alternative options and no economic incentive to diversify. The rapid replacement of traditional agricultural livelihoods with more lucrative tourism-related income sources has dramatically reduced the range of employment options available to people in Khao Lak. Alternative livelihood incomes are supplementary, particularly in the low season when tourist numbers are down to 30 per cent (Calgaro *et al.* 2009: 28). The timing of the disaster, 26 December, meant that the substantial amount of money earned on Christmas Day was washed away by the water because owners had not had the time to deposit the money in the bank (ibid.: 35). An assessment by the Tourism Authority of Thailand (TAT) of the impact on the tourist industry for the first quarter of 2005 resulted in an estimated decrease in the number of foreign visitors to Thailand by approximately five million visitors (Weber 2005: 35).

Earlier work on the 2004 Indian Ocean tsunami

The 2004 Indian Ocean tsunami catastrophe has been surveyed and discussed in a large number of reports and articles. The majority have been about aspects of aid management, security, how to rebuild houses and infrastructure and the environmental impact of the disaster. The academic publications address the Indian Ocean tsunami from different perspectives. However, it is rare that culture and religion are emphasized. An exception is a special issue with the title, *Religions, Natural Hazards, and Disasters* in the journal *Religion*, edited by Jean-Christoph Gaillard and Pauline Texier (2010). Two articles address the 2004 Indian Ocean tsunami: Claudia Merli deals with interpretations of the disaster from Islamic and Buddhist perspectives based on fieldwork in Satun province in Thailand (Merli 2010); I contribute with an article on post-disaster recovery and the role of Buddhist practices in Phang Nga province in Thailand (Falk 2010). I have also published two articles on gender and Buddhism after the tsunami (Falk 2012a, 2012b).

The majority of books published on the tsunami in 2004 are interdisciplinary edited volumes. Matthew Clarke, Ismet Fanany and Sue Keny's (2010) edited volume on the post-disaster reconstruction in Aceh, Indonesia, analyses different perspectives and experiences of the huge post-tsunami reconstruction effort. The book provides a thorough examination of disaster reconstruction and post-conflict issues. The authors explore diverse experiences and assessments of the reconstruction. They undertake an in-depth examination of Aceh as a case study of disaster reconstruction. *Post-Disaster Reconstruction* Clarke *et al.* (2010) consider the importance of the political and religious settings of the reconstruction, the roles of communities and local NGOs (non-governmental organizations), and the challenges faced by Indonesian and international agencies. The authors conclude that reconstruction is not a one-dimensional process; it involves multiple aspects, players, perspectives and experiences. One of the 11 lessons highlighted in the concluding chapter is that the religious and cultural beliefs of the affected community are often central to any reconstruction effort and that an affected community's spirituality is vital and should not be separated from the response to a disaster. Another important lesson drawn from that book is that gender analysis must underscore responses to disasters. The volume *Tsunami and War: Aid, Activism and Reconstruction in Sri Lanka and Aceh* (2009), edited by Malathi de Alwis and Eva-Lotta Hedman, addresses the devastation of the 2004 Indian Ocean tsunami and reconstruction in its aftermath in Aceh and Sri Lanka. The authors come from a variety of disciplines and provide critical analyses of many key categories mobilized in humanitarian discourses and practices. The book contains interesting ethnographic accounts and case studies. The tsunami resulted in a divergence of political trajectories, with Aceh moving towards a cessation of war and Sri Lanka heading towards a resumption of war. These different political outcomes frame this book's compelling arguments about aid, activism and reconstruction. This work questions the frequent disconnection between research and policy and explores ways that this divide can be bridged. However, religion or spirituality is not taken into account in this volume.

Piers Blaikie and Ragnhild Lund edited *The Tsunami of 2004 in Sri Lanka: Impacts and Policy in the Shadow of Civil War* (2009), which is based on empirical research in Sri Lanka. The authors utilize their earlier research experiences of the country in order to learn about the situation after the tsunami. The aim has been to develop new knowledge on post-crisis reconstruction and recovery work. The authors represent various social scientific fields and one goal has been to bridge the knowledge gap between researchers and practitioners. The book deals with housing reconstruction, livelihoods, the internally displaced, humanitarian interventions and prolonged conflicts. Most of the articles in this book are policy-oriented and give important input to recovery work. The volume *The Indian Ocean Tsunami: The Global Response to a Natural Disaster*, edited by Pradyumna P. Karan and Shanmugam P. Subbiah (2011), contains articles written by geographers, geologists, anthropologists and political scientists, and deals with the difficult nature–society relationship after the Indian Ocean tsunami. This interdisciplinary work is divided into two parts. Part one focuses on the environmental and ecological impacts of the tsunami in Thailand, Indonesia, India, the Maldives and Sri Lanka, and part two examines socio-economic dimensions of recovery, reconstruction and response in Thailand, India and Sri Lanka. The book addresses the environmental, social and economic cost of the tsunami and two chapters describe how the tsunami has affected Thailand. One of the chapters gives an informative description of how the Thai environment was affected by the tsunami. The other chapter deals with post-tsunami recovery in Thailand, and focuses on how the tourist industry was impacted by the disaster. Religion is addressed in one chapter, with a case study from India that points out the importance of religious networks and the role of religious institutions in relief work. The volume provides a thought-provoking assessment of the national and international aid responses after the Indian Ocean tsunami, which will provide useful knowledge for those who handle disasters in the future.

Dennis B. McGilvray and Michele R. Gamburd's edited volume *Tsunami Recovery in Sri Lanka: Ethnic and regional dimensions* (2010) explores the post-disaster situation in Sri Lanka from four different disciplinary perspectives: anthropology, demography, political science and disaster studies. This volume stresses the importance of understanding the local cultural context where faith and religion are included in the disaster recovery process. I would like to mention that the introductory chapter contains a selective but interesting survey of earlier work on the Indian Ocean tsunami. A volume that has examined hundreds of reports and articles published on aid assistance after the tsunami is *The Asian Tsunami: Aid and Reconstruction after a Disaster*, edited by Sisira Jayasuriya and Peter McCawley (2010). The authors have investigated the main factors that influenced institutional responses to the delivered assistance following the tsunami disaster, and their aim has been to examine lessons drawn from the aid efforts in order to be useful for disaster protection policies. This volume also stresses the importance of the local cultural context and considers the agency of local communities to be vital in delivering aid, especially during the early phase of a natural disaster.

One of the few monographs dealing with the aftermath of the tsunami is Frida Hastrup's, *Weathering the World: Recovery in the Wake of the Tsunami in a Tamil Fishing Village* (2011). This book is based on anthropological fieldwork in Tamil Nadu, and it explores the various ways in which the tsunami survivors in a fishing village have proceeded to make their local world inhabitable during the years immediately following the disaster. Hastrup provides detailed analyses of restoring the material world, local engagement with the environment, management of local humanitarian support, plans for the future and possible social mobility, as well as commemoration of, and monuments to, the catastrophe. The author demonstrates that the recovery process undertaken by the tsunami survivors aims to simultaneously restore and create something new. It is an interesting and insightful book that focuses on the survivors' agency and the recreation of village life, and displays a notable local recovery process. The book does not, however, address the role of religion in the recovery after the tsunami. *The Golden Wave: Culture and Politics after Sri Lanka's Tsunami Disaster*, by Michele Ruth Gamburd (2014), is an anthropological monograph containing residents' stories of the disaster and its aftermath. It deals with the social, political and economic impact of the tsunami on the Sinhalese villagers' lives and provides valuable insights into how aid and recovery are experienced after a disaster. Finally, *Wave of Destruction: One Thai Village and its Battle with the Tsunami* by the geographer Erich Krauss (2005), is a fascinating and emotional book about one of the worst-hit villages in Phang Nga and some of the people who lived there when the tsunami struck. It provides an interesting description of life in the village before the tsunami and follows some families after the disaster.

The structure of the book

This book is organized into ten chapters and presents experiences and narratives of the catastrophe and the post-tsunami situation from the survivors' perspectives. Chapter 2, "Framing relief and recovery in post-tsunami Thailand", outlines the overall theoretical analytical framework and introduces the reader to the approach of anthropology to disaster. The importance of network is addressed and this chapter aims to contribute to an understanding of conditions for resilience building. The chapter provides a brief overview of how the contested concept of resilience has developed and it introduces religion as a resilience factor. It deals with social capital's positive and negative aspects and also with the interplay between gender, vulnerability and social resilience.

Chapter 3, "The day of disruption", concentrates on the survivors' narratives of how they experienced the disaster and how they survived and dealt with the immediate post-disaster situation. The main focus in this chapter is to show how the survivors felt and acted, and what decisions they made when the tsunami waves struck. This chapter is organized according to where the survivors were when the tsunami hit: some had experiences whilst at sea, some were on land and some were at temples. The narratives are from interviews with Buddhist monks, nuns, male and female survivors and their relatives. In Thailand, as well as in other Asian countries, Buddhist temples are an obvious place to turn to in difficult situations and Chapter 4, "A place of refuge: organizing chaos", explores the activities and the work that was carried out at the Thai Buddhist temples in the tsunami-stricken areas of Phang Nga after the disaster. It deals especially with the first month after the disaster, when the temples were the main places for the identification of dead bodies, the workplace for forensic specialists and sanctuary for thousands of survivors as well as a place to stay for many of the volunteers who came to help. It describes the daily routines and ordinary life at the temples and explains about the relationship between the monks, the nuns and the lay villagers. This chapter also deals with gender and Buddhism and with the role of the nuns after the disaster.

Chapter 5, "Communication across boundaries", explores loss, deaths, funerals and the search for missing persons. It shows how Buddhist monks and nuns were significant in consoling people who had experienced untimely death and ambiguous loss. This chapter investigates how the survivors handled their loss and found that ceremonies, temporary ordination and transference of religious merits to the deceased were of greatest importance. Both Buddhist and local ceremonies are discussed and the chapter examines how rituals and ceremonies became part of the survivors' lives in new ways. They offered psychological reassurance and religious efficaciousness and were the communication links between survivors and the deceased.

Chapter 6, "Buddhism and ways to recovery", focuses on religion's role in the recovery process and how monks communicated with the survivors, answered existential questions and offered people a way of understanding the world. This chapter describes how Buddhist monks and nuns helped the survivors and those affected by the tsunami to find solace in grief. It contains survivors' experiences of the importance of ceremonies, rituals and rites in the recovery process. Encouragement was important, as were religious answers to existential questions. Chapter 7, "Donations, compensations and religious mission", deals with material support through donations and compensations from the Thai government, and how the survivors experienced the relief situation. This chapter also explores the influx of international NGOs and faith organizations that operated in the affected areas after the tsunami. Based on interviews with the survivors, this chapter shows how the conversion of Buddhists to Christianity caused divisions in some villages and created rifts even within families.

Chapter 8, "Remembrance, commemoration and memorial ceremonies", pays attention to memorial practices and the sentiments that memorial commemorations and erected memorials bring about among those affected. It relates survivors' experiences of the tsunami memorial ceremonies and in what ways they became significant in their healing process. It explores some of the divergences between the memorial ceremonies organized locally by the affected communities and the memorial ceremonies arranged by the Thai government. The distinctions occur both in the significance paid to religion and in the politicization of the ceremonies. Chapter 9, "Building resilience", investigates religion as a resilience factor. Findings support the notion that Buddhist explanations, ceremonies and beliefs build resilience. Self-help groups and practical support from the ordained communities were important. The higher level of social capital held by monks in contrast to nuns had an impact. Survivors' social capital influenced their access to support or their exclusion from assistance. The increased participation in Buddhist ceremonies and faith in amulets and ancestral rites were attempts to control an unsure destiny. Survivors lacked trust in the early warning systems, which made them create their own warnings, and long-term loss of livelihood as a result of the tsunami threatened resilience.

Chapter 10 concludes some of the findings and shows that vulnerabilities are dependent upon a complex relationship of various factors, including gender. It demonstrates how culture and religion have the capacity to strengthen resilience building and also how resilience can be undermined and disrupted by actions carried out in the name of religion. This chapter sums up the ways in which Buddhist practices were found to be important in resilience building, both in terms of individual and also communal capacities to prepare for, withstand and adjust to changed conditions after a disaster. Findings show that culture and religious practices develop a sense of connection and wellbeing after a disaster, and religious practices are found to be of special importance in maintaining the feeling of belonging.

Notes

1 This research project is supported financially by the Swedish Research Council, VR.
2 Information from local authorities in Phang Nga, 2005.
3 See www.cbsnews.com/stories/60 20 March 2005 (accessed in April 2006).
4 "Baan Nam Kem: with new power after Tsunami disaster" published by Baan Nam Kem Community Cooperation Centre in July 2006.
5 Interview 3 April 2006.
6 See the Department of Disaster Prevention and Mitigation website – DDPM. www.disaster.go.th (accessed 4 June 2014).
7 The Ministry of Education, August 2005. See www.moe.go.th (accessed on 4 June 2014).

References

Blaikie, Piers and Ragnhild Lund (eds) (2009) "The Tsunami of 2004 in Sri Lanka: Impacts and Policy in the Shadow of Civil War". (Special issue), Norwegian Journal of Geography, 63: 1.

Calgaro, Emma, Kannapa Pongponrat and Sopon Naruchaikusol (2009) *Destination Vulnerability Assessment for Khao Lak, Thailand: Sustainable Recovery and Resilience Building in the Tsunami Affected Region.* Stockholm: Stockholm Environment Institute.

Clarke, Matthew, Ismet Fanany and Sue Kenys (eds) (2010) *Post-Disaster Reconstruction Post-Disaster Reconstruction: Lessons from Aceh.* London and Washington: Earthscan.

Cohen, Erik (ed.) (2008) *Explorations in Thai Tourism: Tourism Social Science Series.* Bingley: Emerald Group Publishing Limited.

de Alwis, Malathi and Eva-Lotta Hedman (eds) (2009) *Tsunami and War: Aid, Activism and Reconstruction in Sri Lanka and Aceh.* New York: South Focus Press.

Gaillard, Jean-Christoph and Pauline Texier (eds) (2010) "Religions, Natural Hazards, and Disasters". *Religion*, 40.

Gamburd, Michele Ruth (2014) *The Golden Wave: Culture and Politics after Sri Lanka's Tsunami Disaster.* Indiana: Indiana University Press.

Falk, Monica Lindberg (2010) "Recovery and Buddhist Practices in the Aftermath of the Tsunami in Southern Thailand". *Religion*, 40: 96–103.

Falk, Monica Lindberg (2012a) "Gender, Buddhism and Social Resilience in the Aftermath of the Tsunami in Thailand". *South East Asia Research*, 2: 175–190.

Falk, Monica Lindberg (2012b) "Merit, Gender and Theravada Buddhist Practices in Times of Crisis". In John S. Harding (ed.) *Studying Buddhism in Practice*. USA and Canada: Routledge, pp. 104–117.

Ferrari, Olivier, Kunlasab Utpuay, Narumon Hinshiranan and Jacques Ivanoff (2006) *Turbulence on Koh Phra Thong.* Paris: Kétos – Anthropologie Maritime, Centre d'Ethno-Technologie en Milieus Zquatiques, Muséum National d'Histoire Naturelle.

Hastrup, Frida (2011) *Weathering the World: Recovery in the Wake of the Tsunami in a Tamil Fishing Village.* New York and Oxford: Berghahn Books.

ISDR (2009) *Risk and Poverty in a Changing Climate: Invest Today for a Safer Tomorrow: Summary and Recommendations.* Global Assessment Report on Disaster Risk Reduction, Geneva: United Nations.

Ivanoff, Jacques (1997) *Sea-Gypsies of the Andaman Sea Post-War Chronicles.* Bangkok: White Lotus Press.

Ivanoff, Jacques (2005) "Introduction and Analysis of the Moken Oral Corpus". In *Moken and Semang: 1936–2004 Persistence and Change*. Bangkok: White Lotus Press, pp. xi–xlv.

Jayasuriya, Sisira and Peter McCawley (eds) (2010) *The Asian Tsunami: Aid and Reconstruction after a Disaster.* Cheltenham, UK and Northampton, MA: The Asian Development Bank Institute and Edward Elgar Publishing.

Karan, Pradyumna P. and Shanmugam P. Subbiah (eds) (2011) *The Indian Ocean Tsunami: The Global Response to a Natural Disaster.* Kentucky: The University Press of Kentucky.

Krauss, Erich (2005) *Wave of Destruction: One Thai Village and its Battle with the Tsunami.* London: Vision Paperbacks.

Lim Kwee Phaik (2011) *Life at Chakrabongse House*. Artarmon: Piccadilly Print.

McGilvray, Dennis B. and Michele R. Gamburd (eds) (2010) *Tsunami Recovery in Sri Lanka: Ethnic and regional dimensions.* London and New York: Routledge.

Merli, Claudia (2010) "Context-bound Islamic theodicies. The Tsunami as Supernatural Retribution vs. Natural Catastrophe in Southern Thailand". *Religion*, 40: 104–111.

Oliver-Smith, Anthony and Susanna M. Hoffman (2002) "Introduction: Why Anthropologists Should Study Disasters". In Hoffman Susanna M. and Anthony Oliver-Smith (eds) *Catastrophe & Culture: The Anthropology of Disaster.* Santa Fe: School of American Research Press.

Phang Nga: 1st Year Tsunami Memorial (2005) Phang Nga/Bangkok: T&D Studio Phang Nga/EPAC Co Ltd.

Umitsu, Masatomo (2011) "The Tsunami Disaster on the Andaman Sea Coast of Thailand". In Karan, Pradyumna P. and Shanmugam P. Subbiah (eds) *The Indian Ocean Tsunami: The Global Response to a Natural Disaster.* Kentucky: The University Press of Kentucky, pp. 35–50.

UNICEF Contribution to Tsunami One Year Later Report, November 2005.

United Nations (2005) *Tsunami Thailand: One Year Later – National Response and the Contribution of International Partners*. Bangkok: United Nations.

Wanaratna, Luecha (ed). (2005) "120 Days After Tsunami Dec. 24, 04: Experiences & Lessons Learnt of Thailand". Thailand: The Ministry of Public Health.

Warunpitikul, Yutthana and Tangwisutijit Nantiya (2005) "Tsunami Aftermath: The Lesson Not Learnt". *The Nation*, 7 March.

Weber, Karl E. (2005) *Tsunami 2004: Nam Chai Thai.* Bangkok: Thai National Identity Board, Royal Thai Government.

2 Framing relief and recovery in post-tsunami Thailand

The day the tsunami struck, 26 December 2004, was a Buddhist holy day in Thailand. Between 9.40 a.m. and 10.40 a.m. local time, the first tsunami wave hit the 954 km-long Andaman Coast. The wave reached Phang Nga province at about 10.30 a.m. At that time, many people had gone to the Buddhist temples to make religious merit by offering food to the monks, participating in the Buddhist ceremonies and listening to the monks' Buddhist teaching. A young Buddhist monk at one of the small village temples close to the sea described his experiences:

> I was at the temple that morning. It was *wan phra* [the weekly Buddhist holy day] and I was in the *sala*[1] together with one *mae chi* [Thai Buddhist nun] and about 30 villagers. We had finished chanting and the laypeople had received the five precepts, when a car full of people passed the temple at high speed.[2] They were screaming and some of the villagers in the *sala* shouted to them and asked why they were in such a hurry. They shouted back, "The water is coming! The water is coming!" Some of the villagers left the *sala* to see what was happening. None of us knew about the tsunami. Other people ran and shouted when they passed the temple but I could not catch what they were saying. For me it was like the end of the world. I could of course have followed the crowd and run but some of the villagers in the *sala* were so frightened and did not know what to do and I felt that I had to stay with them. If I had left them alone, I would have done wrong. It would have been like I had lost my fate and I would have been unhappy for the rest of my life. I thought, if we die, we would die together.
>
> The water rose incredibly fast and we were not safe where we were. We had to move to higher ground and the only possibility that I saw was to climb the trees. I had a very short time to decide which tree would be suitable to climb. I chose the tamarind tree close to my *kuti* [hut]. I started to climb and the villagers followed me. They helped each other up the tree. Some villagers went to the bell tower and some climbed the big Buddha statue and the *stupa* [reliquary monument]. Some villagers tried to escape in a car. When I sat in the tree I saw the car leaving the temple with 19 villagers and shortly after that the water pushed the car all the way to the main road. Only one of the 19 villagers in it survived.

> The water level rapidly became very high and as suddenly as the water had risen, it withdrew again. The water level was alternately high and low for about 20 minutes; and after that when the water ran back to the sea I saw cars, trucks and a lot of things being carried away with the water out to the sea. From my position in the tree I had a good overview and I looked around and saw the huge disaster. Almost all the buildings at the temple were destroyed and all other houses were in ruins. At that point I did not see any dead bodies.
>
> We sat in the tree for more than one hour and when I considered it to be safe the villagers and I climbed down and left the area. The *mae chi* had broken her leg and the villagers brought her to the hospital. I walked to the main road and on the way there were dead bodies everywhere. Most of them were stripped naked by the water. Of the 30 villagers who came to the temple that morning 11 survived the tsunami. I have relatives who live a couple of kilometres from the temple and I walked to them and stayed overnight in their house.[3]

After the tsunami had hit, the Buddhist temples that were not severely damaged became important places for people to seek refuge. Rescue camps were established on temple grounds or just outside, but in close connection with the temples. Temples were also the places where people brought those who had died and they became significant in storing thousands of dead bodies. The Thai people turned to the monks for consolation, for explanations about life and death, and the monks were essential in conducting rites and ceremonies and for restoring continuity and order.

When disasters happen, the strength of a society's entire web of structures and relations is tested. The impact of natural hazards on those who are affected worst and their levels of resilience and chances of recovery have shown not to be at all random (Rigg *et al.* 2008: 138). The tsunami affected people unequally. Poor social and living conditions that were present before the tsunami implied that some were more vulnerable than others. The Oxfam report showed that four times as many women than men were killed by the tsunami in Indonesia, Sri Lanka and India (Oxfam 2005). Among the Thai who perished, 54.8 per cent were female and 45.2 per cent were male (Isalapakdee 2006).

This book examines Buddhism's role in the processes of healing and spiritual beliefs and practices. Access to social capital has been found to be important for a disaster-hit community's recovery. Lacking social capital increases individual's as well as communities' vulnerability in disaster situations (Aldrich 2012). Vulnerabilities are dependent upon a complex relationship of various factors including gender, class ethnicity and age. Social capital's excluding feature might support inequality. Gender orders, socio-economic status and power relations have an impact on how men and women respond to a disaster (see, for example, Enarson and Morrow 1998).

This chapter provides a brief introduction to the development of anthropology in disaster.[4] How quickly people recover after a disaster differs and theories of social capital explain some reasons for the variation. The importance of network and social capital are introduced and aim to contribute to an understanding of conditions for resilience building. However, social capital has a downside of favouring in-groups, thereby making it difficult for those outside the networks and with less social capital to recover (see Aldrich 2012). The interplay between gender, vulnerability and social resilience is addressed and I introduce religion as a factor in developing resilience among people who have gone through one of the most traumatic experiences in life. In order to provide a background to religion as a resilience factor, I provide a brief overview of how the contested concept of resilience has developed. Religion is a force for transformation in difficult times and Pargament (1997) has shown that it can contribute to profound changes in an individual's direction in life.

Anthropology of disaster

Anthropology's contribution to the study of disasters has over the years broadened the focus of disaster research, enabling it to become more contextualized. Anthropology has also added to a deeper knowledge on socio-cultural matters. Anthropological methods, along with long-term ethnographic fieldwork and narratives, give insights into disaster recovery from a local perspective on what affected people need, besides the obvious material necessities. Anthropology of disaster research also provides a deeper understanding of societies' structure, culture, inter-relations and linkages between past and present, which augment the comprehension of how societies are affected (Hoffman and Oliver-Smith 1999). Oliver-Smith states that disasters are deeply embedded in the social structure and culture of a society, as they are in an environment (Oliver-Smith 1999: 25).

Disasters affect the survivors' whole existence and the examination of behaviour and response is an important research focus that anthropology shares with other social sciences (Hoffman and Oliver-Smith 1999: 7). Anthropologists study individuals' and groups' reactions after being hit by disasters and how they handle the post-disaster situation. Issues of risk, vulnerabilities and factors of resilience, are studied by anthropologists and they likewise investigate who frames and defines a disaster (ibid.). Other matters that concern anthropologists are: why a disaster is considered over, who decides that it is over and what happens to the people who have survived a catastrophe when the disaster situation has been declared at an end (ibid.: 9).

Anthropological studies of disaster have in recent decades shown that disasters could be seen as phenomena that occur as something "normal", more or less frequently and sometimes as a result of people living in vulnerable conditions. If people did not live in such vulnerable circumstances, a natural hazard would not be disastrous to people and thereby become a disaster (Oliver-Smith 1996, 1999; Oliver-Smith and Hoffman 2002; Wisner *et al.* 2004; Tan-Mullins *et al.* 2008). Anthropologists have focused on the foundations of vulnerability and also on ways of recovery and have contributed to an understanding of disasters as a process where history, culture and social conditions have an impact both on the direct consequences of a disaster and on the ensuing phases of a disaster, including long-term recovery (see e.g. Oliver-Smith 1999, 2008). How people respond to a natural hazard, how communities struggle for survival and the restoration of hope, and how they deal with risks and vulnerability are central to anthropological research. Inquiries about behaviour, relations and access to resources and networks, are important for the whole process of recovery. This includes ways of adapting and possibilities for people to make choices about their lives after having lost family members, homes and livelihoods in a disaster.

Anthropological studies on disasters are found in special journal issues, in edited volumes and sometimes also in reports. Two recent anthropological monographs on post-tsunami recovery are Frida Hastrup's *Weathering the World: Recovery in the Wake of the Tsunami in a Tamil Fishing Village* (2011) and Michele Ruth Gamburd's *The Golden Wave: Culture and Politics after Sri Lanka's Tsunami Disaster* (2014). Hastrup explores the recovery process in Tamil Nadu and gives an interesting ethnographic account of the villagers' experiences of the tsunami; she argues that, over time, the disaster became interwoven into ordinary life. Gamburd has, over many years, conducted research in Sri Lanka and she followed the consequences of aid for local people on the southwest coast of Sri Lanka after the tsunami. Her book examines power relations after the catastrophe. She collected stories about the tsunami and shows how the disaster has changed social identities, economic dynamics and political structures in Sri Lanka.

Disasters put societies and cultures to the test because how societies function prior to natural hazards are crucial for the outcome of the post-disaster recovery. Hoffman and Oliver-Smith (1999: 10) point out that anthropology has been central in stressing the importance of research into pre-disaster conditions and the significance of analysing how these influence recovery from disaster, as well as the impact of disaster on different groups in society, because the same disaster affects people differently. Vulnerability, for example, is multi-dimensional and the result of political, ecological and economic relations that people have with the environment (see Oliver-Smith 2008: 11). Not all poor people are vulnerable to disaster, but vulnerability and poverty are social conditions that have been found to mutually reinforce each other (Delica-Willison and Willison 2008: 146–147). In Thailand, poverty has stark regional features. Levels of poverty are highest in outlying regions and lowest in Bangkok and the surrounding areas.[5] From a regional perspective, poverty is not predominant in the south of Thailand. However, the provincial variations in living standards are even starker than the regional differences and the differences within provinces on sub-district levels are striking (see Healy and Jitsuchon 2007). Provinces with low poverty rates such as Phuket exist alongside provinces with higher poverty rates like Ranong and Phang Nga.

Vulnerability focuses on limitations or lack of access to resources and it is often the result of inequality. Oliver-Smith states that: “the concept of social vulnerability expresses the multidimensionality of disasters by focusing attention on the totality of relationships in a given social situation which constitute a condition that, in combination with environmental forces, produces a disaster" (Oliver-Smith 2008: 11). Examples of social vulnerability are unequal relationships that exclude certain groups or individuals from benefiting from disaster risk reduction or post-disaster recovery efforts. Such power relationships manifest themselves between individuals or socio-economic groups, and within institutional frameworks. Anthropological research has long contributed to an understanding of exposed people's vulnerability and anthropology could add to a refinement of the nuanced concept of social resilience.

Vulnerability is commonly discussed in relation to social resilience and they are sometimes viewed as opposites. Manyena points out that resilience and vulnerability are exclusive concepts and it is important to identify and map resilience to show that vulnerability is not contrary to resilience. As with groups of people or communities, individuals can also be resilient in certain situations and not in others. Furthermore, individuals can be resilient in non-resilient communities and there are non-resilient individuals in highly resilient communities (Manyena 2006: 444–446).

Vulnerability contains a combination of external and internal aspects. Some people run a higher risk of being affected by disasters and some people have more difficulties than others coping and recovering after a disaster. Social vulnerability has been shown to be most apparent in the immediate wake of disasters, when unequal patterns of suffering and recovery gain attention. In all countries that were hit by the Indian Ocean tsunami, more women than men perished (Oxfam 2005). A report, *Destination Vulnerability Assessment for Khao Lak, Thailand* (Calgaro *et al.* 2009) reveals how vulnerable the local villagers in the worst hit tourist area in Phang Nga had become because of their total dependence on tourism-related incomes. Before the area became an attractive tourist destination, livelihoods were diverse and that changed when the profitable tourism business was introduced in Phang Nga. Diversification disappeared and, as a result, the villagers lost a tried-and-tested strategy for reducing vulnerability and building resilience against difficulties (ibid.).

Anthropologists have long taken an interest in examining suffering, and this theme includes research on how suffering is embodied and how experiences of suffering can have transformational consequences for individuals and groups. Families' and societies' changes in the different phases of suffering, relief and recovery after disasters have been portrayed (see, for example, Oliver-Smith and Hoffman 1999; Zaman 1999; Kleinman *et al.* 1997). Dealing with suffering is central to the Theravada Buddhist teaching, and this book aims to add to our knowledge about Buddhism and disaster recovery by providing findings of how Buddhist monks and laypeople deal with suffering in post-disaster recovery.

Gender and disaster

Women and men are vulnerable to disasters in different ways and these are often based on economic and social orders (see Phillips and Morrow 2008; Enarson and Chakrabarti 2009). Disaster recovery processes validate a society's pre-disaster gender order and disasters reveal inequalities in the societies that are affected. Disasters occur in gendered contexts, and in disasters women are generally defined as more vulnerable than men (Fordham 2008: 176–178). Pre-existing unequal gender relations make women poorer and more vulnerable than men. In most societies, women are economically weaker compared to men and an unequal economic position, together with an institutional subordination in society, leaves women more vulnerable than men. Such order excludes women from taking part in rebuilding their society after a disaster and, thereby, they lack influence over their own future. Studies have shown that women are affected worse than men when a disaster strikes. Women also run a high risk of receiving fewer benefits than men when recovery begins (Valdés 2009: 18). The Asia Pacific Forum on Women, Law and Development (APWLD) has noted that, based on women's subordination in society, women commonly do not participate in those settings that plan rebuilding after a disaster.

Disasters impact men and women differently in terms of survival, death, injury, trauma and recovery (Ariyabandu 2009: 14). Following the Indian Ocean tsunami, women accounted for as much as 80 per cent of the dead in some parts of Indonesia, India and Sri Lanka (ibid.: 11). It has been explained that the greater number of female deaths in the tsunami mirrors gendered facts, such as the fact that most women in the affected countries have not been given the opportunity to learn to swim or climb trees. Female dresses, fashioned from ankle-length cloths, make it impossible to move rapidly. Women tried to find their children before escaping from the tsunami. They carried children and helped elders who needed their assistance, and that made them more vulnerable (ibid.: 11–12).

It is notable that gender issues are commonly overlooked in disaster situations, especially since gender relations are so prevalent in daily life and in society at large. Also, gender in disaster research has long been neglected. Enarson states that it is first during the last two decades that focus on gender has developed into an important field in disaster research (Enarson 2004: 71). In the beginning of the 1990s, international relief organizations pioneered research by introducing gender issues into their training programmes. Slowly, studies with gender focus have become part of disaster research (ibid.: 52–58). One exception is South Asia, where non-governmental organizations were quite prompt in actively addressing and reporting on women's vulnerability in disaster situations. In other parts of Asia, gender-focused writings have, until recently, been scarce. Zenaida G. Delica wrote one of the first accounts on women's vulnerabilities in South-East Asia in 1998. Her research deals with the Mount Pinatubo eruption in the Philippines (Delica 1998).

Enarson recognizes principal differences in ways of dealing with gender issues between researchers from affluent nations and researchers from disaster-vulnerable, low-income nations in the developing world. Researchers from rich countries focus on the individual in the disaster situation and researchers from developing countries have a more encompassing approach in understanding disasters and pay attention to history, culture, politics, economy and the environment (Enarson 2004: 62–64). Some reasons for this could be that – post-disaster – most studies by researchers from wealthy countries are driven by the disastrous event. Moreover, they conduct short-term studies that often lack historical perspectives, where the aim is probably not to analyse primary causes.

Ariyabandu has found that actions that are supposed to be gender "neutral" in reality force women into situations of distress. Instead of receiving protection and relief aid, women risk becoming victims of violence and their opportunities to obtain relief aid are much less than the opportunities available to men (Ariyabandu 2009: 15). One example from the post-tsunami situation in Thailand are widows who did not receive compensation for lost fishing boats, based on a preconceived belief that it was the husbands who were the fishermen and not the wives. It is however, common for husband and wife to fish together. If fishing was the family's livelihood, the widow could still earn a living from a fishing boat by hiring fishermen without fishing herself.

In Thailand, the Buddhist temples are places of refuge in times of difficulty and laypeople visit monks and nuns for counselling. Buddhism in Thailand is gendered throughout and it is only men (monks and novices) that constitute the monastic community. At some temples the Thai Buddhist nuns (*mae chi*s) are welcome to live in a separate section, but they are not considered to be members of the *sangha*. Some *mae chi*s live in "nunneries", which are governed by the *mae chi*s themselves (see Chapter 4). This gendered Thai Buddhism had implications for how monks and nuns could assist the survivors.

Social capital and disaster research

Generally, social capital refers to the way that individuals and groups invest in and share resources in order to gain return. Social capital is related to social support, formal and informal social ties, organization linkages and cooperation, citizen participation and leadership and roles. Social capital is also related to attachment to a place and a sense of community, and potentially affects the impact of disasters as well as economic development, information and communication, and community competence (Norris *et al.* 2007). Social capital could be understood as social resources that have dynamic qualities and transformational characteristics. Social capital is thus comprised of network structures, linkages and the social support that the network provides, which comes from family, friends, neighbours and colleagues, as well as agencies and institutions. Social capital also encompasses relationships between individuals and their wider neighbourhoods and communities (ibid.: 20).

There is an extensive scholarship on social capital not connected to disaster research, and social capital is defined in different ways and used in various disciplines. Pierre Bourdieu, James S. Coleman and Robert D. Putnam are renowned for having developed theories of social capital. Pierre Bourdieu introduced several forms of capital. He defines social capital as resources that are actually or potentially linked to memberships within groups (Bourdieu 1986:

248). Bourdieu includes both collective resources and networks that facilitate access to social capital. His analysis of social capital revealed it helped individuals to position themselves and build social hierarchies. According to Bourdieu, the size of a person's network is important, including the magnitude of the capital that other network members have, and to which the individual gains access (ibid.: 252). Bourdieu, Coleman and Putnam see social capital as a resource for individuals and society.

Coleman described outcomes on the basis of the types of relations that exist between individuals within families and communities (Coleman 1988). Putnam defines social capital as "connections among individuals – social networks and the norms of reciprocity and trustworthiness that arise from them" (Putnam 2000: 19). For Putnam, social capital involves relations of reciprocity and trust that enable people to solve their problems, and he recognizes both the individual and the collective side of social capital (ibid.: 20).

Social capital could broadly be described as either focusing on the resources linked to the possessions that memberships in networks bring (Bourdieu 1986) or on features of social organization where trust, norms and networks can coordinate actions and improve the efficiency of society (Putnam 1993). The different types of social capital are characterized as bonding, bridging and linking social networks. Bonding social capital refers to relations that are similar to us while bridging social capital signifies ties to people or groups unlike ourselves. Bridging social capital is commonly more difficult to build than bonding social capital. Usually, bonding and bridging social capital connect individuals of the same status, while linking social capital connects networks of trusting relationships that have different status and authority in society (Aldrich 2012: 33). Bonding social capital includes personal network relations with, for example, family, friends, neighbours and colleagues as well as impersonal network associations with agencies and institutions (Norris *et al.* 2007: 20). There are findings in research on social capital and disaster that state that bonding social capital – or close networks – is important for immediate support, but bridging and linking social capital offer support for longer-term survival and wider neighbourhood and community recovery (Hawkins and Maurer 2010). Social capital has also been found to increase financial and material donations (Baker and Refsgaard, 2007) as well as lobbying activities for disaster response (Nakagawa and Shaw, 2004).

A common definition of social capital is "resources embedded in social relations and social networks" (Lin 2008: 4). Nan Lin argues that the importance of social capital lies not in structures and network but in resources accessible through one's "direct and indirect ties" (ibid.: 4). According to Lin, the best source of social capital is a relationship that is personal, informal and lasts over time. In contrast, a relationship that is public, formal or short-lived is a poorer source of social capital (ibid.: 83).

Social capital has been proven to help people during and immediately after a disaster has struck. This is apparent before the government has mobilized and sent resources, and researchers agree that it is in the local community that individuals help each other to survive and that here, social capital plays a vital role (see Aldrich 2012; Baker and Refsgaard 2007; Moore *et al.* 2004; Nakagawa and Shaw 2004). It is a common understanding in the disaster management literature that communities with higher levels of social capital are able to fare better when faced with an emergency (Murphy 2007: 302). Daniel Aldrich adds that it is usually assumed that the speed of recovery following a disaster will be determined primarily by the extent of the initial damage and economic conditions, but recent research has begun to challenge this assumption (Aldrich, 2011). He even argues that higher levels of social capital are more important than greater economic resources, assistance from the government or outside agencies, and low levels of damage when it comes to facilitating recovery and helping survivors coordinate more effective reconstruction (Aldrich 2012: 2). Norris supports Aldrich's statement and asserts that there are few studies that relate social capital to the disaster cycle. Those studies that have applied the concept to the disaster context observe that social capital may be a prerequisite for an adaptive capacity promoting resilience, where loose but cooperative community ties are the most effective for rapid response and recovery (Norris *et al.* 2007), and that social capital increases participation in and speed of recovery (Nakagawa and Shaw 2004).

Putnam recognizes that "faith communities in which people worship together are arguably the single most important repository of social capital in America" (Putnam 2000: 66), but that acknowledgment of religious networks' importance for social capital has not implied that religious associations have been especially recognized, and religion's unique role in building social capital has been given little attention (Smidt 2003: 2). Examining religious associations and social capital would render knowledge about social activities as services and resources to their members and others in the community, for example, by providing care, social support and social networks.

Social capital is considered to be important for resilience (see Aldrich 2012). In this study, I focus on social capital from a network and inequality perspective inspired by the work of Aldrich (2012) on social capital and resilience building and by the work of Pargament (1997, 2010) on religion and resilience building. In Thailand the Buddhist ordained life is strictly regulated and creates a sharp boundary between the ordained community and the laypeople. Being ordained implies that the person has withdrawn from the lay life and no longer belongs to their natal family; instead, they have been adopted and incorporated by the ordained community. At temples, life is organized so members of the *sangha* have access to both bonding and bridging social capital. The high degree of social capital in the *sangha* is built on trust and reciprocity. The social capital that arises becomes crucial for mobilizing the community to provide aid programmes or participate in other forms of activity.

Downside of social capital

Social capital has proved to be helpful to many but not all, and equity issues are not always taken into account in social capital literature. Scholars have examined both the positive and the negative outcomes of social capital, but the preferred emphasis has been on the positive effects. The tendency to overlook the drawbacks that social capital also creates has been criticized. Aldrich has found that "peripheral or marginalized groups within society that hold less social capital benefit little and often are harmed by the groups holding stronger social capital after a disaster" (Aldrich 2012: 14). Social capital can be used to gain power over others. Strong links can also lead to a group failing to network with those who are different from them. It has been shown that the real strength of resilience is in the level of inter-community links between various groups (Murphy 2007).

Women are described as commonly having less social capital than men, especially in contexts with stronger gender stratification systems (Lin 2008: 7). Maxine Molyneux (2002) points out that there are considerable differences between men's and women's networks and inclusion and exclusion are almost always gendered processes. She concludes that power relations within societies are reflected in and reproduced by social networks (Molyneux 2002: 181). Molyneux has examined social capital in development policy thinking and practice in Latin America and states that we need to observe the fact that women are very often central to the forms of social capital that development agencies and governments are keen to mobilize in their development programmes (ibid.: 167). I would argue that it is likewise important to keep this fact in mind for disaster research.

In Thailand as well as in other countries with large low-income groups, ties to family and kin are strong and become a refuge in times of difficulties. Many of those Thai people who lived in the small communities that were struck by the tsunami were originally from other parts of Thailand. Therefore, they did not have their network of family members and relatives in the affected provinces. I found that it was common for people who had migrated from the same province to choose to become neighbours in their new village and form informal networks as a substitute for close relatives. Informants from Phang Nga told me that they engaged in reciprocal supportive relations and this seemed to be important to both men and women.

Religion is an integrated part of many Thai people's lives. However, more women than men regularly participate in local forms of associational life. More women than men were represented in voluntary self-help groups concerning, for example, health, education, food and livelihoods. These supported new networks and social capital, which sustained the social fabric of the community. Both positive and negative sides of social capital were present in post-tsunami Thailand and this will be addressed in following chapters.

Defining social resilience

Alexander (1997) and Hartman and Squires (2006) argue that natural disasters are fundamentally social. Research has shown that the impact of a natural disaster is the product of how the society is organized (Wisner *et al.* 2004, Tan-Mullins *et al.* 2008: 116). The lack of equality in the social, political, economic and religious arenas, as well as which individuals or groups have access to resources and which have not, becomes clearly visible after a disaster.

The concept of resilience has developed over time and has also been adopted by the social sciences. Resilience is used in several disciplines and it has been questioned whether resilience is relevant to both natural and human systems (Manyena 2006). Consequently, definitions of resilience are manifold and sometimes resilience is used interchangeably with coping and is viewed as the opposite to vulnerability. Resilience both communicates the capacity to survive sudden disturbances and is also a measure of the capacity to recover from the impact of a hazardous event. In early definitions of resilience, the concept was seen as more reactive and outcome-oriented. Those definitions focused on restoring the situation as soon as possible to what it was before the disaster occurred. Later definitions have shifted to become more process-oriented and involve capacity building, mitigation, emergency preparedness, planning changes and development of the affected community, and through these means disaster resilience can lead to a desired outcome and development (ibid.: 437–439).

It is interesting to note that when resilience was first studied in the 1940s, it was within the disciplines of psychology and psychiatry (Manyena 2006: 233). Based on the fact that individuals can be resilient in non-resilient societies, psychiatrists and psychologists investigated the resilience of individuals and their interaction in society. About three decades later, in the 1970s, the concept appeared in the fields of ecology and physics. Holling published his influential work *Resilience and Stability of Ecological Systems* in 1973. Ecologists explored the concept by studying resilience in organisms and life systems. In the 1980s, the political scientist Aaron Wildavsky dealt with the cultural theory of risk and was one of the first researchers to introduce the resilience concept in the social sciences. In the social sciences, the concept of resilience has been broadly defined in terms of individuals' or communities' adaptive capacities following difficulties or hardships. Resilience is seen as the capacity to cope with unanticipated dangers and includes the ability to recover and continue to function (see Wildavsky 1991; Kendra and Wachtendorf, 2003). Considering the varying meanings of resilience, the concept implies sturdiness or resistance in tandem with elasticity and flexibility.

Social resilience is an elusive and complex concept and a broad definition of social resilience is the capability of social groups and communities to recover from or respond positively to crises (Maguire and Hagan 2007: 16). In the context of disaster, social resilience is expressed in the commitment of various segments of society to join forces for the achievement of common goals, in their ability to cope with threats over extended periods of time and in their ability to adapt to changes. This designation of social resilience contains a motivational component that refers to individuals' willingness to contribute to the common good and to help the collective achieve its goals, and another component refers to the ability to cope with actual or potential threats while maintaining a reasonable level of orderly functioning.

The United Nations' International Strategy for Disaster Reduction (2005) makes use of resilience in "The Hyogo Framework for Action 2005–2015:

Building the Resilience of Nations and Communities to Disaster". In this strategy plan, resilience is defined as:

> the capacity of a system, community or society potentially exposed to hazards to adapt, by resisting or changing in order to reach and maintain an acceptable level of functioning and structure. This is determined by the degree to which the social system is capable of organizing itself to increase this capacity for learning from past disasters for better future protection and to improve risk reduction measures.
>
> (UN/ISDR 2005: 4)

This definition implies that disaster-affected societies will be supported in order to become resilient and more attention will be paid to what the affected societies themselves can carry out and how they can build on their own resources.

Mileti signifies local resiliency with regard to disasters by the ability to withstand an extreme natural event without suffering devastating losses, damage, diminished productivity or quality of life, without a large amount of assistance from outside the community (Mileti 1999). However, communities are complicated and research has shown that they can be resilient to some disasters and vulnerable to others (Maguire and Hagan 2007: 19). Moreover, different groups within one society can be more or less resilient based on access to social capital. Theories of human resilience often refer to adaptation in the face of stress or adversity. Human resilience may be interpreted as a multi-dimensional, open and fluid process of reaction and response. Resilience is both a common and a culturally specific process and could be seen as a cultural practice, a local and everyday experience directed toward human security and sustainability that emerges within and in relation to current and specific conditions.

Religion in relief and recovery work

A disaster creates questions of different kinds and cultural and religious themes are often employed to address existential questions and to give death a meaning. Religion provides answers to core questions about the meaning of life, death and suffering and the psychologist Kenneth Pargament has written extensively about how religious beliefs and practices contribute greatly to resilience (see, for example, Pargament 2010). Religion has proved to be a resilience factor in disaster situations and different religious doctrines may affect the ways in which people view human nature generally; they are of special importance in post-disaster recovery. Religious practices, such as prayer, are also closely linked to resilience (Mitchell 2003). Prayer for protection from disaster is accepted by most faiths as appropriate, provided that prayer is matched with positive human action to ensure the protection of persons and property (ibid.: 25). Buddhists do not believe in God and have no God to turn to for protection. Protection should be created by individuals themselves, built by their own skilful actions based on not violating precepts and good relations with other people. Local monks told me that after the tsunami the ceremonies that laypeople asked for served three purposes: (1) Transferring religious merit (*bun*) and creating better circumstances for those who were killed by the tsunami, (2) communicating with ghosts and informing them that they are dead and must leave and not create trouble for the living, and (3) making religious merits for the survivors' families, relatives, friends and all living beings, including themselves. Local religious practices become part of a wider scope of religious activity in diverse situations of disaster. Religious revitalization and hybridization could be interpreted as ways for individuals to establish protection and security and build resilience (see Chapter 9).

Buddhism's core values were relevant for most of the Thai Buddhist survivors. In Buddhism, all natural occurrences are accounted for in terms of the Laws of Dependent Originations (*Paticcasamuppada*). The laws signify that everything is interconnected and everything is relative and conditional; what happens at one moment is determined by what preceded it. Karma (*kamma*) is central to Buddhist beliefs and karma means morally relevant actions – intentional actions that result in future states of being. In Thai usage, these are the consequences of morally relevant actions of the past, including past lives, and explanations as to why some died and some survived the tsunami were commonly attributed to the persons' karmic status. Foreign tourists who had survived the tsunami frequently recognized Thai people's emphasis on generosity and service to others. The basis for generosity and helpfulness in the Thai culture and Buddhism is understood to create beneficial karma and is something that Thai people learn from a very young age. Religious behaviour may contribute to social capital formation in that volunteering, charitable contributions and other distinct acts of mercy can, at least in the short run, help to provide a "safety net" for members of society who are "at risk" (Seagert *et al.* 2001: 215).

Social inequalities have an effect on how people access and use social capital following a hazard and they influence access to networks, information and resources. The Buddhist *sangha*'s strong position and high level of social capital was evident in post-disaster Thailand. The monk who featured at the beginning of this chapter belongs to the *sangha* and to the extremely well developed national and international Buddhist network. He survived the tsunami and on the same day of the disaster he decided to walk to his relatives, who lived in the region. His temple was in ruins and more temples nearby were damaged. The monk told me that immediately after the tsunami, he did not consider returning to the place of the temple and he did not think of rebuilding it.[6] However, he changed his mind when his lay followers, monks and the *sangha* invited him to come back and continue his duties as a monk in the temple that was destroyed. Buddhist networks raised money and rebuilt the temple and it has developed into a larger temple than it was before the tsunami. The strength of the *sangha*'s social capital is manifest. The Thai monks at that temple cooperate with monks from Myanmar and the Myanmar monks conduct ceremonies at the temple in the Myanmar language for the migrants from Myanmar who work in the area. Laypeople from both countries visit and support the temple. This temple has enlarged its social capital by also including Buddhist nuns. Even before the tsunami, this temple accepted nuns and it is one of the few temples in the tsunami-struck Phang Nga province that host nuns.

Recovery is defined in a number of ways and most of the definitions of post-disaster recovery focus on physical recovery; less attention is paid to social recovery. As Nakagawa and Shaw point out: "disaster recovery is not only about building houses but reconstruction of the whole community as a safer place" (Nakagawa and Shaw 2004: 12). Research has linked higher levels of social capital to superior health outcomes, stronger governance and better economic growth (Aldrich 2012: 13). Aldrich argues that social resources can assist in recovery by serving as informal insurance after a disaster, overcoming collective action problems that hinder recovery and rehabilitation, thus strengthening "voice" and decreasing the probability of exit (ibid.: 14). Strengthening "voice" means that people speak up about their needs and "exit" denotes leaving the area. Survivors with strong social networks experience faster recoveries and have access to required information, tools and assistance. Simultaneously, those who have weak social capital have difficulties coping (ibid.: 15).

In this study resilience is explored through survivors' subjective experiences of their recovery. I have found that Buddhist memorials and merit-making rituals have the capacity to strengthen and support the survivors. In the stressful post-tsunami period, the Buddhist temples kept on performing the daily practices with ceremonies and meditation in spite of all the tasks that the monks had to carry out. The following chapters address the ways in which temples became important for the recovery process and the significance of Buddhist ceremonies, memorial services and other activities run by monks for bonding within the affected community, which enabled members to share their loss and commemorate the deaths of loved ones together.

This book addresses the high level of social capital among the monks and the *sangha* in contrast to the lower social capital among the nuns, which was particularly evident in the absence of temples in the affected areas that could accommodate nuns. Female informants expressed that they had trust in female Buddhist leadership, but they did not have access to counselling with nuns. Many laywomen had high levels of bonding social capital and they used social capital in the process of resilience building. During the post-tsunami recovery, considerable interfaith dialogue among most Buddhist groups in Thailand reflected their high levels of bridging social capital. Co-religious activities were included in the annual commemoration of the tsunami and these activities were organized nationally and also by the local government.

Concluding remarks

The Thai government responded rapidly to the tsunami disaster and the Thai Ministry of Public Health (MOPH) mobilized quickly. The first team from Bangkok arrived in Phuket six hours after the tsunami struck and in the first week of January 2005, about 90,000 persons in affected communities, temporary shelters and so on had received medical and mental health care (MMWR 2005). In the early stages of the recovery process, a large number of NGOs and volunteers provided immediate relief and resolved urgent problems. However, for a long time after, the tsunami the relief operations were chaotic and sometimes dysfunctional (see, for example, Tan-Mullins *et al.* 2008: 123–125, and Chapter 7, this monograph).

In a disaster of the magnitude of the Indian Ocean tsunami, most of the affected countries needed local, national and international assistance. Commonly, national and international relief organizations emerge at the affected site after a disaster. However, the best institutions and persons to distribute relief are found in the locality and that calls for cooperation. Anthropologists and disaster researchers in other fields have shown that relief aid allocated by local institutions and leaders risks being distributed unevenly and there is a danger that distribution of aid leads to corruption and develops power relations (Hoffman and Oliver-Smith 1999: 11).

In Thailand, the more affluent province of Phuket was shown to be much more resilient than the less prosperous provinces of Ranong and Phang Nga. Phuket's greater social capital increased the recovery speed and the Thai government initially concentrated its relief efforts on Phuket (see e.g. Calgaro *et al.* 2009). Ranong and Phang Nga present several so-called pockets of poverty, especially in the coastal areas where the tsunami hit (Healy and Jitsuchon 2007). Informants complained that most rebuilding efforts were focused on Phuket's tourist industry and many in Phang Nga and Ranong felt overlooked. Informants were disappointed and said that the government had promised a great deal but had not fulfilled its assurances.[7]

In this book, I aim to contribute to a broader understanding of ritual, religion and resilience in disaster situations. This anthropological study explores survivors' experiences of the disaster and covers individual monks' and nuns' efforts to assist the survivors in coming to terms with their experiences, install confidence and encourage them to not give up but continue living. Throughout this book I examine post-disaster recovery with focus on how Buddhist monks, nuns and the temples supported the survivors after the tsunami and in what ways religion was a resource for survivors in the recovery process. The Buddhist monks and nuns used their social networks to access information and exchange favours. Their cooperation with laypeople, other Buddhist groups and organizations was essential in the relief work. The Buddhist temples and networks were important in assisting the local temples and the survivors in building resilience in post-disaster recovery. In contrast, religious activities such as missions and conversion in exchange for relief aid are tragic examples of how the Thai survivors became exploited in vulnerable situations (Falk 2013). This book addresses the intersection of gender, ways of recovery and religious practices. Employing an anthropological approach to Buddhist post-tsunami practices implies taking into account modes of action that are gendered and highly socially and contextually influenced. In the following chapters I seek to map out the role of Buddhism in the recovery process.

Notes

1 A *sala* is a building that is open to both laity and monastics, where ceremonies and other gatherings take place.
2 The five precepts: to abstain from killing, stealing, sexual misconduct, lying and taking intoxicants.
3 Interview, 15 April 2007.
4 In this chapter, passages are reproduced from Falk (2012) "Gender, Buddhism and Social Resilience in the Aftermath of the Tsunami in Thailand", published in *South East Asia Research*, 2: 175–190, copyright SOAS (Falk 2012).
5 Over several decades Thailand experienced a rapid economic growth, and from 1986 to 1996 the Thai economy had an average annual growth of 9.1 per cent. By 1996, aggregate poverty had fallen from 29.5 to 11.4 per cent. Poverty increased to 15.9 per cent after the economic crises in 1999 (Healy and Jitsuchon 2007: 758).
6 Interview 15 February 2007.
7 The survivors repeated that comment over the years that I conducted fieldwork.

References

Aldrich, Daniel P. (2011) "Between Market and State: Direction in Social Science Research on Disaster". *Perspectives on Politics*. 9(1): 61–68.

Aldrich, Daniel P. (2012) *Building Resilience: Social Capital in Post-Disaster Recovery*. Chicago and London: The University of Chicago Press.

Alexander, David (1997) "The Study of Natural Disasters, 1977–1997: Some Reflections on a Changing Field Of Knowledge". *Disasters*, 21(4): 284–304.

Ariyabandu, Malalgoda Madhavi (2009) "Sex, Gender and Gender Relations in Disasters". In Elaine Enarson and P.G. Dhar Chakrabarti (eds) *Women, Gender and Disaster: Global Issues and Initiatives*. New Delhi, California, London and Singapore: Sage Publications, pp. 4–18.

Baker, Daniel and Karen Refsgaard (2007) "Institutional Development and Scale Matching in Disaster Response Management". *Ecological Economics*, 63(2–3): 331–343.

Bankoff, Greg, Georg Frerks and Dorothea Hillhorst (eds) (2008) [2004] *Mapping Vulnerability: Disasters, Development, and People*, London: Earthscan.

Bourdieu, Pierre (1986) "The Forms of Capital". In John Richardson, ed., *Handbook of Theory and Research for the Sociology of Education*. New York: Greenwood Press, pp. 241–258.

Calgaro, Emma, Kannapa Pongponrat and Sopon Naruchaikusol (2009) *Destination Vulnerability Assessment for Khao Lak, Thailand: Sustainable Recovery and Resilience Building in the Tsunami Affected Region*. Stockholm: Stockholm Environment Institute.

Coleman, James S. (1988) "Social Capital in the Creation of Human Capital". *American Journal of Sociology*, 94: 95–120.

Delica, G. Zenaida (1998) "Balancing Vulnerability and Capacity: Women and Children in the Philippines". In Elaine Enarson and Betty Hearn Morrow (eds) *The Gendered Terrain of Disaster: Through Women's Eyes*. Westport, CT: Praeger Publishers, pp. 109–113.

Delica-Willison, Zenaida and Robin Willison 2008 [2004]. "Vulnerability Reduction: A Task for the Vulnerable People Themselves". In Greg Bankoff, Georg Frerks and Dorothea Hillhorst (eds) *Mapping Vulnerability: Disasters, Development, and People*. London: Earthscan, pp. 145–158.

Enarson, Elaine (2004) "International Perspectives on Gender and Disaster: Differences and possibilities". *International Journal of Sociology and Social Policy*, 24 (10/11): 49–93.

Enarson, Elaine and P.G. Dhar Chakrabarti (2009) *Women, Gender and Disaster: Global Issues and Initiatives.* New Delhi, California, London and Singapore: Sage Publications.

Enarson, Elaine and Betty Hearn Morrow (eds) (1998) *The Gendered Terrain of Disaster: Through Women's Eyes.* Westport, CT and London: Praeger.

Falk, Monica Lindberg (2012) "Gender, Buddhism and Social Resilience in the Aftermath of the Tsunami in Thailand". *South East Asia Research*, 2: 175–190.

Falk, Monica Lindberg (2013) "Thai Buddhists' Encounters with International Relief Work in Post-Tsunami Thailand". In Kawanami, Hiroko and Geoffrey Samuel (eds) *Buddhism, International Relief Work, and Civil Society.* New York: Palgrave Macmillan, pp. 27–49.

Fordham, Maureen (2008) [2004]. "Gendered Vulnerability Analysis. Towards a More Nuanced Approach". In Greg Bankoff, Georg Frerks and Dorothea Hilhorst (eds) *Mapping Vulnerability: Disasters, Development, and People.* London: Earthscan.

Gamburd, Michele Ruth (2014) *The Golden Wave: Culture and Politics after Sri Lanka's Tsunami Disaster.* Bloomington: Indiana University Press.

Hartman, Chester and Gregory D. Squires (2006) *There is no Such Thing as a Natural Disaster: Race, Class, and Hurricane Katrina.* New York and London: Routledge.

Hastrup, Frida (2011) *Weathering the World: Recovery in the Wake of the Tsunami in a Tamil Fishing Village.* New York and Oxford: Berghahn Books.

Hawkins, Robert and Katherine Maurer (2010) "Bonding, Bridging, and Linking: How Social Capital Operated in New Orleans following Hurricane Katrina". *British Journal of Social Work*, 40(6): 177–1793.

Healy, Andrew J. and Somchai Jitsuchon (2007) "Finding the poor in Thailand". *Journal of Asian Economics*, 18(5): 739–759.

Hoffman, Susanne and Anthony Oliver-Smith (1999) "Anthropology and the Angry Earth: An Overview". In Anthony Oliver-Smith and Susanna Hoffman (eds) *The Angry Earth: Disaster in Anthropological Perspective.* New York and London: Routledge.

Holling, Crawford, Stanley (1973) "Resilience and Stability of Ecological Systems". *Annual Review of Ecology and Systematics*, 4: 1–23.

Isalapakdee, Pimolwan (2006) "Tsunami: Deaths and Injuries in the Affected Areas". Unpublished, presentation at the 2nd Annual Conference of Population and Social Research, Asia Hotel, Bangkok. Organized by Research Institute of Population and Social Research, Mahidol University (in Thai.)

Kendra, James M. and Tricia Wachtendorf (2003) "Elements of Resilience after the World Trade Centre disasters". *Disasters*, 27(1): 37–53.

Kleinman, Arthur, Veena Das and Margaret M. Lock (1997) *Social Suffering.* Berkely, CA: University of California Press.

Lin, Nan (2008) "A Network Theory of Social Capital". In Dario Castiglione, Jan W. van Deth and Guglielmo Wolleb (eds) *The Handbook of Social Capital.* New York: Oxford University Press, pp. 50–69.

Maguire, Birgit and Patrick Hagan (2007) "Disasters and cCommunities: Understanding Social Resilience". *The Australian Journal of Emergency Management*, 22 (2): 16–20.

Manyena, Siambabala Bernard (2006) "The Concept of Resilience Revisited". *Disasters*, 30(4): 433–450.

Mileti, Dennis S. (1999) *Disaster by Design: A Reassessment of Natural Hazards in the United States.* Washington, DC: Joseph Henry Press.

Mitchell, Jerry T. 2003. "Prayer in Disaster: Case Study of Christian Clergy". *Natural Hazards Review*, 4(1): 20–26).

MMWR (2005) "Rapid Health Response, Assessment and Surveillance After a Tsunami – Thailand, 2004–2005". *Morbidity and Mortality Weekly Report*, 54: 61–64.

Molyneux, Maxine (2002) "Gender and the Silences of Social Capital: Lessons from Latin America". *Development and Change*, 33(2): 167–188.

Moore, S., M. Danniel, L. Linnan, M. Campbell, S. Benedict and A Meier (2004) "After Hurricane Floyd Passed: Investigating the Social Determinants of Disaster Preparedness and Recovery". *Family & Community Health*, 27(3): 204–217.

Murphy, Brenda L. (2007) "Locating Social Capital in Resilient Community-level Emergency Management". *Natural Hazards*, 41: 297–315.

Nakagawa, Yuko and Rajib Shaw (2004) "Social Capital: A Missing Link to Disaster Recovery". *International Journal of Mass Emergencies and Disasters*, 22(1): 5–34.

Norris, Fran H., Susan P. Stevens, Betty Pfefferbaum, Karen F. Wyche and Rose L. Pfefferbaum (2007) "Community Resilience as a Metaphor, Theory, Set of Capacities, and Strategy for Disaster Readiness". *American Journal of Community Psychology*, 41(1–2): 127–150.

Oliver-Smith, Anthony (1996) "Anthropological Research on Hazards and Disasters". *Annual Review of Anthropology*, 25: 303–328.

Oliver-Smith, Anthony (1999) "'What is a Disaster?': Anthropological Perspectives on a Persistent Question". In Anthony Oliver-Smith and Susanna Hoffman (eds) *The Angry Earth: Disaster in Anthropological Perspective.* New York and London: Routledge, pp. 18–34.

Oliver-Smith, Anthony (2008) [2004]. "Theorizing Vulnerability in a Globalized World: A Political Ecological Perspective". In Greg Bankoff, Georg Frerks and Dorothea Hillhorst (eds) *Mapping Vulnerability: Disasters, Development, and People*. London: Earthscan, pp. 10–24.

Oliver-Smith, Anthony and Susanna M. Hoffman (eds) (1999) *The Angry Earth: Disaster in Anthropological Perspective.* New York and London: Routledge.

Oliver-Smith, Anthony and Susanna M. Hoffman (2002) "Introduction: Why Anthropologists Should Study Disasters". In Hoffman Susanna M. and Anthony Oliver-Smith (eds) *Catastrophe & Culture: The Anthropology of Disaster*. Santa Fe, NM: School of American Research Press.

Oxfam (2005) "The Tsunami's Impact on Women". *OXFAM Briefing Note* (March).

Pargament, Kenneth I. (1997) *The Psychology of Religion and Coping.* New York: Guilford Press.

Pargament, Kenneth I. and Jeremy Cummings (2010) "Anchored by Faith: Religion as a Resilience Factor". In John W. Reich, Alex J. Zautra and John Stuart Hall (eds) *Handbook of Adult Resilience.* New York: Guilford Press, pp. 193–210.

Phillips, Brenda D. and Betty Hearn Morrow (2008) *Women and Disasters: From Theory to Practice.* Philadelphia, PA: Xlibris.

Putnam, Robert D. (1993) *Making Democracy Work: Civic Traditions in Modern Italy.* Princeton: Princeton University Press.

Putnam, Robert D. (2000) *Bowling Alone: The Collapse and Revival of American Community.* New York: Simon & Schuster.

Rigg, Jonathan, May Tan-Mullins and Carl Grundy-Warr (2008) "Grounding a Natural Disaster: Thailand and the 2004 Tsunami". *Asia Pacific Viewpoint* 49(2): 137–154.

Saegert, Susan, J. Philip Thompson and Mark R. Warren (eds) (2001) *Social Capital and Poor Communities.* New York: Russell Sage Foundation.

Smidt, Corwin (ed.) (2003) *Religion as Social Capital: Producing the Common Good.* Waco, TX: Baylor University Press.

Tan-Mullins, May, Jonathan Rigg and Carl Grundy-Warr (2008) "Responses and Resilience of Fisherfolks on the Tsunami Event in Southern Thailand". In Timothu Doyle and Melissa Risely (eds) *Crucible for Survival: Environmental Security and Justice in the Indian Ocean Region.* New Jersey, New York and London: Rutgers University Press, pp. 116–129.

UN/ISDR. 2005. www.unisdr.org/wcdr/. (Last accessed 7 June 2014).

Valdes, Helena Molin (2009) "A Gender Perspective on Disaster Risk Reduction". In Elaine Enarson and P.G. Dhar Chakrabarti (eds) *Women, Gender and Disaster: Global Issues and Initiatives.* New Delhi, California, London and Singapore: Sage Publications, pp. 18–28.

Wildavsky, Aaron (1991) *Searching for Safety*. New Brunswick, NJ: Transaction Press.

Wisner, Ben, Piers Blaikie, Terry Cannon and Ian Davis (2004) *At Risk: Natural Hazards, People's Vulnerability and Disasters*, 2nd edition. London and New York: Routledge.

Zaman, Mohammad Q. (1999) "Vulnerability, Disaster, and Survival in Bangladesh: Three Case Studies". In Anthony Oliver-Smith and Susanne Hoffman (eds) *The Angry Earth: Disaster in Anthropologocial Perspective*. London: Routledge, pp. 192–212.

3 The day of disruption

Experiencing a natural disaster and the loss of close family members and relatives has forever affected the lives of the people involved. This chapter concentrates on the victims' narratives of how they survived and experienced the tsunami and the immediate post-disaster situation. The main purpose is to show how the survivors encountered the tragedy, how they felt and acted, and what decisions they made when the waves smashed. The tsunami hit the Thai shores on a Sunday morning when the schools were closed and many children were at home watching a popular children's programme on TV. That Sunday coincided with the weekly Buddhist holy day (*wan phra*) and some people, especially women, were at the temples offering food to the monastics, receiving the Buddhist precepts from the monks and listening to their Buddhist teachings. The day of the disater, 26 December 2004, was described as a beautiful day and perfect for fishing and therefore many were at sea when the waves struck.

Most of the temples (*wat*s) in the tsunami area are small and have about eight to ten resident monks. Several of the *samnak song*s, small Buddhist "temples" that lack an ordination hall and do not have full accreditation as a *wat*, have only a few resident monks. The day that the tsunami occurred was a full moon day. The monks and laypeople call it *wan phra yai* (the 'big *wan phra*'), which occurs monthly. *Wan phra* is the Buddhist observance day, which follows the lunar calendar and takes place once a week. In Thailand, monks, and also some nuns, go on alms rounds every morning and once a week, on *wan phra*, laypeople also bring food to the temple. Most Theravada Buddhist monks eat twice a day but there are also monks, belonging to the Thammayut sect, who only have one meal a day. Monks who have two meals a day have their first meal in the early morning at about 7 a.m. and the second before 12 noon. According to their monastic discipline, no Theravada monks or nuns are allowed to eat after noon. Donating food to the ordained community is considered meritorious and a part of many Thai Buddhists' daily lives. At the temples on *wan phra*, after the laypeople have donated food to the monks, the monks give Buddhist speeches, conduct chanting and thereafter give the five Buddhist precepts to the laypeople.

This chapter is organized according to where the survivors were when the tsunami hit: some had experiences whilst at sea, some were on land and some were at temples. The monks' narratives are from temples that were struck by the tsunami and they share their thoughts and describe how they handled the situation. Most of the narratives are from interviews with Buddhist monks and male and female survivors and their relatives; they were collected between January 2005 and January 2008.

At sea

Wimon and Noi

Wimon was 46 years old when the tsunami hit the island where he was born and had lived all his life.[1] Wimon was Buddhist and he is married to Noi, who was Muslim and born in the province Nakhorn Si Thammarat. Before they married, Wimon converted to Islam. He said that this was expected of him according to the Muslim convention. Later, however, both Wimon and his wife converted to Buddhism and today he, his wife and their two children are Buddhists. Wimon said that they decided to become Buddhists because of their occupation. According to Wimon, most Muslims work on plantations, but he had always been a fisherman and it was therefore better for them to be Buddhists. Noi said that she had nothing against converting to Buddhism. The conversion would probably have been problematic if she was very young, said Wimon, but Noi was old enough to make her own decision. Wimon and his wife had only four years of primary education. However, they considered education important and both their daughter and son have university degrees. Their children had been away from home studying and now their son worked in Bangkok and their daughter worked in Phuket and they were therefore not on the island on the day of the tsunami.

Wimon and his wife were at sea when the tsunami reached where they were fishing. Wimon was busy trying to get the squid traps from the sea when his wife saw the tsunami wave approaching at high speed. She shouted out that something looking like a rocket or a missile was coming towards them. Wimon immediately dropped the squid traps and accelerated as fast as possible towards an island. He said that he could not predict the direction of the tsunami wave and therefore it took time to escape. The tsunami wave was different from what he called a natural wave. He said a natural wave follows a line but the tsunami wave changed direction.

Wimon and Noi saw two small fishing boats that were closer to the island than they were. Suddenly, the boats strangely disappeared. They listened to the radio and heard news from Nam Khem that an island close by had disappeared under water and was then totally destroyed. At first they did not believe what they had heard and thought it was a joke. Wimon called their friends on Surin island and they told him that the same had happen to parts of Surin. This made them believe that what they had heard was true and they understood that the situation was serious. In that moment Wimon changed his mind and instead of heading towards the island, which he initially considered to be the safest way to escape the troubled sea, he turned the boat and went as fast as he could out to deep sea. They said that decision probably saved their lives.

They had some empty barrels in the boat and Wimon tied them to his wife's body so she would float if the boat sank. Noi was crying and said that she was terrified. They had never seen anything similar to this before. They described an earlier experience when they were fishing near the Burmese border and the Burmese began shooting at them. Noi was very afraid when that happened, but she said that was nothing compared to experiencing the tsunami.

Wimon had been fishing for 30 years, but he had never seen waves as big as the tsunami waves. He thought that the boat would be smashed into pieces if it were directly hit by such waves. Further out to sea, they succeeded in escaping the waves and they stayed there until the sea became calm. When the gigantic waves had ceased, they tried to use the phone but it was dead. They spotted a speedboat and Wimon took that as a sign that it was possible that not all places were affected.

They decided to return home to the island of Kho Phra Thong. When they reached Village Three on the island, they saw that more boats had gathered and the fishermen suggested that all the boats should travel together. The sea was full of things floating around. When they reached the big canal through the mangrove at Village Three and Village Four almost 30 boats that had survived the tsunami headed towards Village Four. Some fishermen recognized items that belonged to them floating in the water. Wimon said he felt disheartened when he saw all the objects in the water. To begin with, he did not see anything that he thought would be from their house but suddenly he recognized the frames from their windows. Their house was the only one that had that type of windows. When Noi saw the frames she started to cry. Before they had reached Village Four, they received a call on the radio from the village leader who told them to come to the mainland and asked the boats not to go to Village Four, because that part of the island was completely destroyed and there was nothing left. Wimon was about to steer the boat to Kura Buri but Noi persuaded him to continue to Village Three and see their new house. They had slept only one night in the house and she wanted to see with her own eyes what was left of it.

The house had been completed about two-and-a-half months before the tsunami hit, but they had continued living in the old house because they had not yet arranged the house warming ceremony with the monks. They had planned to have the ceremony when their daughter was back home to celebrate New Year. When the boat reached the tip of the island they could see that every building was demolished; not a single building was standing and the sea and land were full of rubbish. They got the impression that the island had changed because they could not even recognize the place where their house had been. In the wake of the tsunami, there were after-shocks and the tide was still alternating between high and low. Therefore, they did not dare to leave the boat. Happy to be alive but devastated about losing their home, they went to Kura Buri on the mainland. They said their boat seemed to be the last boat that arrived from the sea to Kura Buri that day. Wimon's relatives and sister were waiting for them at the pier. His sister invited them to stay at her house and she served them food. Wimon was exhausted and said that he could not eat a bite.

Mina

Mina lived on the mainland south of Koh Phra Thong.[2] When the tsunami struck she was 43 years old, married and a mother of three children – two sons and one daughter, and several of her relatives were her closest neighbours. They lived in what had developed into a prominent fishing village in Takua Pa district. It was a small fishing village close to the Andaman Sea and Mina's house was located very near the shore. They had their fishing equipment outside the house and it was walking distance to where they had their fishing boat.

On Sunday 26 December, Mina's eldest son and her daughter were not at school. That morning, her youngest son was flying kites together with his friends on the beach not far from the house. Usually, Mina and her husband did not go out fishing together. Her husband normally went fishing with employed fishermen; however, the hired fishermen had a couple of days off work. That morning Mina told her husband that she wanted to go out fishing so she could give her mother some fresh fish to take back home with her to Nakhorn Si Thammarat province, where she lived. That particular day, Mina had to persuade her husband to go with her out to sea. Her husband said that he did not feel well and wished to stay at home and rest. Mina was persistent about going because her mother appreciated the fish she was given. Nakhorn Si Thammarat province is several hours away by bus and her mother had come to help them look after their youngest son while Mina and her husband were working. Her mother had wanted to return home in the early morning on 26 December, but she agreed to stay one more day, since Mina had arranged for her to travel by car the next day and that was more comfortable than by bus.

Mina described how they prepared for the fishing trip early that Sunday morning. She took a new set of nets to replace the older ones and she hoped to catch a lot of fish. They left the house and Mina's mother, their two sons and daughter at home. Mina and her husband went far away out to sea before they cast out the fishing nets. Her older brother, who lived next door, had also gone fishing that morning. However, he did not go far and was fishing close to the coast. His wife, daughter, a granddaughter and a cousin of his wife were at home.

Mina said that fishing at sea that morning was the same as usual until they suddenly felt that the sea was acting strangely and, unexpectedly, there were big waves. They managed to control their boat through the waves. They had never experienced anything like that before and felt that it was dangerous. They decided to leave the nets in the sea and try to return to the pier as soon as possible. When they came closer to the mainland they were terrified to see that the sea was full of floating objects and dead people. Mina said she became very worried about her family at home. On their way to the pier they rescued some naval soldiers who were struggling in the water after their ship had sunk. The soldiers told them about the big disaster at Tablamu Navy Base and that made them even more worried about the situation at home.

When they arrived at Tablamu pier, close to the naval base, they saw the devastation. As soon as they reached land they left the boat and all the fishing gear. Tablamu was not close to their village, so they tried to get a car to take them back home. Mina said that it was almost impossible to find a car and people told them that the road was closed. Eventually, they found a car and a driver who was willing to take them to Takua Pa district.

At 10.38 a.m., the first wave had reached Mina's village and crashed into their house. Her mother was at home, their oldest son was visiting some friends in another village, their daughter had gone to watch the sea that was acting so bizarrely before the tsunami hit and their youngest son was at the beach with his friends playing with his kite. As soon as Mina and her husband reached their village, they wanted to rush to their house and search for their family. They found that everything in their village was in ruins and it was not possible to get to their house. They tried desperately, without success, to find their family members. Later the same day they found their daughter and oldest son alive at the district office in town where they had been taken after they were rescued, but they did not find their youngest son.

Charon

South of the fishing village where Mina resided, and north of the tourist destination Khao Lak, lived Charon and his wife.[3] He was about 45 years old when the tsunami struck. He used to be a fisherman and had his own boat and employed fishermen. Charon was born in that area. His father died when he was two years old and his mother died when he was eight. He grew up together with his siblings and was taken care of by his older brothers and sisters. His siblings also stayed in the area where they were born and they worked on a rubber plantation. Charon finished school after grade seven. Before he began fishing, he worked for a couple of years in the tin mines in Nam Khem. While working in Nam Khem he lived at the mining camp. When the number of people increased and Nam Khem became crowded, Charon left because he wanted to live in a less busy place. He kept in touch with his friends in Nam Khem after he left the village and he used to visit them to have a chat and relax after coming back from the sea.

For many years, Charon fished for squid but three years before the tsunami struck he had an accident when he was working at sea. He hurt his back badly and the chronic backache he suffered from after the accident made it impossible to carry out the heavy work of a fisherman. Some of Charon's friends had begun to work in the tourist business and they invited Charon to take tourists on boat trips. He had kept his boat and thought that could be a good way to earn some money, so he accepted the offer. He adjusted the boat to suit his new occupation and equipped it with life jackets for the tourists. The hotels at Pakarang booked the boat tours, prepared lists of tourists and told Charon where they wanted to go. Charon's wife said that he was popular among the tourists and he always tried to make the trips enjoyable for them.

On 26 December 2004, Charon was at sea with a group of eight tourists. When the tsunami hit, they were in shallow water. He decided to drive the boat towards the canal and told the tourists to run as fast as they could. He said he did not know that it was more dangerous on land than at sea. The land was flat and there was no hill to climb. The moment he and the tourists had left the boat it was totally destroyed by a wave in front of their eyes. Charon said that the tsunami sounded like a plane taking off. A German lady fell down and he had to help her on to her feet again. He could not think of how they could escape; he said he did not understand what was happening, so he just ran as fast as he could. He said that the water came from two directions – from the back and from the front. The water that came from the beach was black in colour and was not at all transparent; it looked dirty. The waves were very big and he said he noticed that they split into smaller waves. He saw that the waves had taken boats and they were floating on the water out of control.

Charon said that he realized that where he had landed the boat was not a safe place. If he were to survive he had to fight the tsunami. At home, the other villagers could escape by running to higher ground but he was stranded on the flat land and his boat was totally destroyed. When the tsunami was over, he and the tourists could not do anything but wait and hope for a boat to come and pick them up. But there were no boats. He and the tourists decided to divide themselves into two groups and walk in opposite directions and try to look for boats that were passing by. Suddenly, Charon heard the sound of a big boat. The tourists did their best by waving to the boat, but nobody saw them. Charon then ran to get a better position and finally the people on the boat saw him. They were taken on board and slept there one night. All of them were in shock and the children and women were crying.

Charon's wife was at home when the tsunami hit and she took refuge on the hill. She was certain that her husband had died. She knew where he had taken the tourists and that area was badly affected by the tsunami. Other people who had been at sea and survived were back, but Charon was still missing. Both Charon and his wife considered it a miracle that he and all the tourists who were with him survived. Charon told about the importance of local knowledge of where to escape. He described a man from his village who saved many people. The man was travelling in a bus when he saw the tsunami coming. He stopped the bus, got off and shouted to the other passengers to follow him when he ran up the mountain. He knew the way and those who understood and trusted him followed and they survived.

Charon concluded that the persons who had stayed on the road died at once and most of the villagers who perished in his village had been close to the sea when the tsunami struck. Also, those who had knowledge about where to go could nonetheless not escape if they were close to the sea. His nephew tried to run away from the tsunami. He was at work and went to the sea to see what was happening. The sea had withdrawn about 3 km. His nephew started his motorcycle and went as fast as he could when he saw that the boats were turned over. He stopped and shouted to his friends to run away. When he reached the main road, there was a car accident and he had to find another route. He saw many people who were run over by trucks. People were running and the drivers wanted to escape and thus did not stop; they just drove on. Some died because landlords had built fences that blocked the road. Some fences were so high that it was impossible to climb over them. Charon said that another nephew died while he was trying to escape the waves and ran into an area with a high fence. More people could have survived if such fences had not blocked the escape routes.

Tom and Nit

Tom had lived in Ranong province for five years when the tsunami hit.[4] He was 40 years old, married to Nit and they had three children. They were both born in southern Thailand – Tom in the province of Trang and Nit in Phang Nga. Tom had finished seventh grade and had been a fisherman for about 25 years. He learned the profession by working together with his father on his father's boat. After that he was employed as a fisherman on big fishing boats and later when he and his wife moved to Ranong, he bought his own long-tail boat. Both Tom and Nit were born Buddhists and they lived in a village with both Muslims and Buddhists. He said that life was quite uncomplicated before the tsunami. They had everything they needed in the house and his fishing equipment was complete. Everything changed with the tsunami, and he said that after that they had nothing. The house was gone. The fishing gear was gone. The motorcycle was gone.

When the tsunami hit the coast of Ranong, Tom was at sea. His wife Nit and their three children were at home. Usually, he and Nit went out fishing together but that day she was not with him because he was not fishing at that time. He had been asked by some people in the village to help look for dead bodies because there was a criminal who had killed and dumped several victims in the sea. Tom had originally planned to take the day off and stay at home, something he seldom does, but when the villagers asked him to search for the bodies he said he was, of course, willing to help. The persons that the criminal had murdered were Buddhists and therefore, he explained, they could not carry out the search with a Muslim boat. It must be done with Buddhist boats. Tom said that they had not found any of the dead bodies when the tsunami struck them.

Tom went with his boat and there were three more boats that carried out the search. Because it was a Sunday, the rest of the family remained at home. From Monday to Friday their two older children, a boy and a girl, were at school. While he and his wife went fishing, a woman who lived not far from them further down the road looked after their youngest child. Tom and Nit were fishing almost every day and on Saturdays and Sundays when their children were home from school, their son looked after his two younger sisters.

Tom summarized with a smile that they were going to look for dead bodies in the sea and while doing so they risked becoming dead bodies themselves. There were, altogether, three persons in Tom's boat. In each one of the other three boats there were three or four people. Among the search team, only one person died in the tsunami, even though all four boats were destroyed. Some people managed to swim to a nearby island. Tom's boat was robust and had a good engine and it could therefore make it to the island despite the damage. On the island they waited for somebody to come and pick them up, but nobody came to their rescue. They decided to try to go in Tom's boat and in one of the other boats even though the boats were badly damaged. On their way back, Tom helped two tourists, who were floating in the sea, on board his boat. Tom and the other men reached the mainland at about 5 p.m. that day.

They had heard on the radio about the big waves and that the area where they lived was totally destroyed. When they reached the place that used to be their village, all the buildings were gone. They found only dead bodies. The first thing that all of them did was to run to their houses and look for their families. In order to find his house, Tom looked for a special tree that was close to his home and sadly he found that the house was completely demolished. He said there was nothing left.

Tom searched for his family but he could not find them. He went to the hospital and there he found his wife and later also their son; to his relief both had survived. He continued looking for the other two children. The first night after the tsunami, Tom, his wife and his son slept at the hospital. Tom said he could not sleep at all because he was full of worry about their missing daughters.

In Ranong province most of the deceased were brought to the hospital. At the hospital the next day, Tom found one of his missing daughters who had died in the tsunami. He went to the Buddhist temple Wat Kamphuan to make a vow. There were organizations that had put together research teams for localizing dead bodies. However, Tom said that they did not look in the difficult-to-reach places. Therefore, the villagers also formed groups of four-to-five people that searched for missing family members. Tom and Nit did not have any family or relatives in the area, so Tom went by himself to search for his other daughter. He said the search was very difficult because the water had brought boats, motorbikes, cars and other things, and the corpses were often under those heavy objects, which made searching extremely hard. However, he found his daughter's body in the mangrove on the first day of his search. He said that he was totally devastated. They had lost two daughters, aged just eight and four.

Tom compared his situation with persons who were not directly affected by the disaster through the loss of family members and he said that it is probably easier for them to recover. For him it was a very serious matter and he found it difficult to heal. He also found it hard to talk about his feelings and not having any relatives in the area made it worse.

To begin with they lived in a tent and later they acquired a house in one of the new settlements that have been built for the survivors. In the new settlement the majority are Muslims, but there are also some Buddhist families. Tom said they decided to live there because it is close to the shore and easy to go fishing and it is also easy for their son to go to school from there.

On land

Duang and Ann

Duang and Ann are two fishers in their fifties who lived in a fishing village in Takua Pa district.[5] They were both Buddhists and Duang especially was particular about conducting – and participating in – religious ceremonies. Ann was born in the province of Prachuap Khirikan, but had lived almost her whole life in Takua Pa together with many of her relatives who were still her closest neighbours. Duang's parents died when she was a child. She was first brought up by relatives in Phuket province and later adopted by a family in Takua Pa.

Duang described how they had returned from the sea one hour before the tsunami hit. Duang had driven off on her motorbike to sell their catch of crabs and fish to fishmongers' shops located between the fishing village and the main road. Ann was at home and had begun preparing crab-traps with bait for the next fishing trip. Duang recounted that she had sold all the crabs but had not yet sold any fish when she heard about the big waves that were hitting the village. She left the fish and went on her motorbike as fast as she could back to her house. She could not reach the house with her bike and so she left the motorbike on the hill not far from the village. She said she was very worried about Ann and about Ann's father, who lived with them.

On her way back home she met people running from the village and they were screaming that everything was destroyed. At first, Duang could not believe them. Then she met a woman who was covered in dirt and her hair was full of mud. When she saw her she understood that something terrible had happened. Some people who fled were running completely naked. Some of Ann's relatives were running to escape the water; she asked them if they had seen Ann and her father and they answered that both were at home.

When she reached their house she found Ann, on the roof with her father. Duang said that Ann's father was more than 100 years old and not very mobile. Ann had, with help from one of her friends, carried the old man up on to the roof. Duang came to the house after the second wave had hit and the house was filled with water. She said that the water reached up to her chest. She could manage to walk by catching and holding on to the electric cables. She climbed up on to the crowded roof with about 30 people and she looked around at the other houses and saw that many Burmese workers were sitting on the roofs. The area where they lived was not so badly affected as it was not close to the shore, but the water in the canal was turbulent.

Ann thought that it would be safer for them to be in the boat. She tried to get down from the roof and into the boat but she could not because there were too many people on the roof and it was difficult to move. Duang said that they wanted to take the boat and drive through the canal to a safe place, because the water in the canal seemed to be fairly calm. However, it turned out later that their boat had been destroyed and carried away by the waves.

All three sat on the roof until they thought it was safe to go down. The father was sent to the hospital in a rescue car and Ann and Duang left the village by foot. Duang said that she saw dead bodies everywhere right after the tsunami. She stated that what she saw was unreal. She thought that the dead children looked like floating dolls and she found it strange that most dead bodies were black in colour. Some corpses were in a sitting position, while others looked like they were moving. The bodies had been battered by wood and smashed by heavy objects and they were trapped under all the objects and materials that the tsunami had brought with it. Two of Ann's relatives were missing after the tsunami and their bodies were never found.

Ann's older sister lived next door to them and she was in Songkla province attending a funeral when the tsunami happened. She said that she had left Songkla at about 8 a.m. in the morning. At about 10.30 a.m., she received a call from her daughter who told her about the tsunami, and Ann's sister said that she did not know what a tsunami was. Her daughter was crying and told her that she was staying at the temple. Then the line suddenly went dead and it was impossible for Ann's sister to get in touch with anybody by phone. She travelled back home in a car driven by her son. She asked the police what had happened but they did not know. She also listened to the radio, but there was no information or news about the catastrophe. It was between 1 and 2 p.m. in the afternoon that she first heard some news on the radio. They said that a tsunami had hit Takua Pa, Nam Khem and Kao Lak in Phang Nga province. When she heard that, she told her son to drive faster. At about 3 p.m. she received a call from her son who was in Takua Pa and he told her that his grandfather and their relatives were safe. All had managed to escape the waves by staying on the roof. She said that there was a tamarind tree in front of her house, and that particular tree had saved many of the villagers' lives.

Duang and Ann went to the hospital in the afternoon to visit Ann's father. They asked a nurse where he was, but the nurse was crying and could not answer. They asked a doctor, but he was also crying. In the end they were all crying because the situation was chaotic and the hospital was so overcrowded. Eventually they found Ann's father lying on the floor in a room with many patients. That same day, they made arrangements for him to be transferred to a hospital in Surat Thani province. He stayed in Surat Thani a couple of weeks and then wanted to stay with Ann and Duang in Takua Pa. They had not finished cleaning the house, so he stayed with them in a house that they had rented. The house was not comfortable to live in and it proved difficult for Ann's father, who had problems walking. When they had cleaned their house, they moved back and the father stayed with them until April. He was then admitted to Takua Pa hospital where he later died.

Cleaning the house was time consuming. Duang and Ann began cleaning about a week after the tsunami had hit and it took them several weeks to make the house habitable again. Duang said that there was a very unpleasant and indescribable smell everywhere in the village. The rescue teams were working hard wrapping and carrying away corpses. Helicopters were also used to take dead bodies from the mangrove forest. Their house was located opposite the mangrove forest and the waves brought many victims there. Duang said that they could see across the canal what the corpses looked like and she also mentioned that there were thousands of flies. The flies were big and had green heads. She said she was surprised that there were not any dead bodies in the area around their house. A long time after the tsunami, villagers who were fishing crabs in the mangrove forest said that there were still corpses in the forest and they had found skulls there. Duang explained that the mangrove forest is dense and it was hard to reach the deceased that were deep inside the area. She said that she experienced how very difficult it was to walk in the forest when she participated in the mangrove reforestation programme after the tsunami.

Niwat

Niwat and his wife were Buddhists and they lived in a village where the majority of people were Muslims.[6] He had been married twice. He had three children from his first marriage and no children from the second. Niwat was born in Petchaburi province, his father was a fisherman and he had chosen to follow the same occupation. In Petchaburi he was an employed fisherman, but the job he had was not a steady one. He moved to the south to seek a more secure livelihood. He settled in Phang Nga and he managed to realize what he had hoped for – buying his own fishing boat. He said he liked fishing much more than gardening because fishing offered him independence and he preferred working without supervision.

In the morning of 26 December 2004, his wife had listened to the news on the radio and heard about an earthquake in Indonesia. They did not know what a tsunami was and there were no warnings on the radio about a possible tsunami with giant waves. Earthquakes happen frequently in South-East Asia, without any consequences outside the earthquake area and therefore they did not pay much attention to the news. Niwat was at home because he never went fishing on *wan phra*. That morning, he and his wife were carrying out the usual work they did at home: his wife was repairing the fishing nets and he was repairing the fishing gear.

Niwat explained that the area along the canal was divided into a southern and northern section. The tsunami hit the southern part first and people from that area came running and shouting at the top of their lungs that a huge wave was on its way. Niwat said that their warnings saved their lives, even if they at first did not believe that the waves could be so dangerous. The people from the southern area were running in different directions. Some were looking for relatives and some people ran to escape the waves.

Niwat and his wife left the house and Niwat stayed so he could have a good look at what was happening. He was worried about his two-storey house and his possessions. He told how the tsunami came with the booming sound of an aeroplane. His house was directly struck by the tsunami and was smashed into pieces. Niwat pointed out the direction of the tsunami along the canal and said that it was like a domino effect. All the houses along the shore fell one by one.

Niwat said that there was only 5 m between him and the wave, and the water level rose rapidly. The wave destroyed everything in his house and also his boat and fishing equipment. He and his wife were not hit by the wave and they survived without any injuries. But, to their sadness, they lost Niwat's son and his grandson in the tsunami. His son had his own family and lived in a house close to the water. They had been at home when the tsunami came and the wave had taken them by surprise.

Niwat did not leave the affected area until that afternoon. The government officers helped the survivors to move to the village school. Niwat remembered that the authorities said that a tsunami was going to strike again, but that did not happen. Living in the school was, according to Niwat, like living in a refugee camp and both Buddhists and Muslims were placed there. There is a Buddhist temple not so far away, but all people from the affected area where Niwat and his wife lived stayed at the school. The school was the centre for the distribution of food and other necessities. People who lived far from the sea and also people from other provinces came to help out. Niwat did not recall how many days he stayed at that school; he thought it was probably about ten days. After that, he went back to the location of his house. He explained that in his district there were no temporary shelters like there were in Takua Pa. Those who had lost their houses stayed in tents and people who had houses that were not destroyed moved back to live in them.

Niwat said that everything they had was gone and they did not even have the basic necessities to sustain their lives. The boat was destroyed and Niwat had thereby lost his livelihood. The place where his house had been built was extremely dirty and they had to clean up the area. He said that they had to be careful about how they used the water because it was contaminated with disease.

Those affected by the tsunami waited for some help from the government. They had heard that the government gave compensation to people in Phuket and they had already begun to rebuild what was destroyed. While waiting for support, Niwat worked as a daily worker, hired by the government to clean up the area and collect the refuse, and he earned 180 baht per day (about US$22).

Niwat mentioned that they could use pieces of wood from their old house but clothes and other things were not possible to use. Before the tsunami, Niwat had built a shack close to the water because it was convenient to have a storage place and basic shelter when he was fishing. He thought that it was very strange that his simple shack was spared and not destroyed by the tsunami.

Niwat said that the best way to tackle what had happened was to accept the situation and to adjust to it (*tham jai*). He declared that the tsunami is a natural phenomenon and it was not a malevolent being who wanted to cause them trouble. Niwat's wife said that after the tsunami, Niwat was concerned about how they should survive. He talked about the fishing equipment, about not having any money, not having a house and not having a boat. These worries made him suffer (*thuk*). They had heard that those who lost property in the tsunami would receive donations, and they waited for that. Niwat said that if he had not received a boat he would have moved back to Petchaburi, because without a boat they could not stay in Phang Nga; he would then have become jobless. They found it unthinkable to start over again in Phang Nga with two empty hands. The couple said that they could live a very simple life and they could actually live anywhere, including in the shack. Niwat added that if he had fishing equipment, he could get a new house. The loss of his son and his grandson was tragic but Niwat considered losing family members as something that was commonplace (*thammada*).

After the tsunami the government provided dried food, instant noodles and canned sardines to the survivors. Niwat and his wife lived on that kind of food for many months until he received a boat from an international NGO. Later, the government provided Niwat and his wife with a small house in a new settlement. Niwat said that it took a long time to carry out a survey and draw up an inventory of what the families had lost. The government did not allow people to build houses in the area where the destroyed properties had been. Niwat would have preferred to have the house at the same place. However, most people who had lost their property wanted to build houses on the new land provided by the government. The old land was actually part of a national park and not private land, but they had lived there without problems for decades. In order to avoid complications and the risk of creating conflicts by making special requests, Niwat and his wife agreed to build a house in the settlement together with the other villagers. However, he said the location of his old house was much more convenient for fishing and he did not feel comfortable living close together with other people in the new house in the settlement.

Yen and Daeng

Yen and Daeng are a married couple without children who originally came from the provinces of Sakhon Nakhorn and Petchaburi.[7] They moved to the fishing village in Takua Pa and Daeng was employed as a fisherman on boats. Later, they bought a boat and Daeng and his wife fished for several years together. Before the tsunami the couple rented a house near a canal. On the day of the tsunami they had come back from fishing and, having had sold their squid, were preparing for the coming two days of fishing.

Both of them were making squid-traps outside the house when the tsunami hit. Yen noticed that the canal became totally dry, which was very strange and something she had never seen before. When the water filled the canal again it was full of bubbles. When she told her husband about the phenomenon, he said that they should run. She thought that it would have been safer for them to stay in the boat. They saw the wave rolling in and they ran as fast as they could. Yen said if they had waited just a short while they would probably have been pushed into the mangrove. Yen wore a sarong and her husband wore short trousers. They ran and they shouted to people they passed that they should also run, but some did not believe them. She bumped into a Burmese person when she was running and fell, and her husband helped her up again. They said that it was lucky that neither of them were badly injured.

They ran quite far away from the village. There were rumours about where it was safe to be. The temples were very crowded. A car picked them up and they ended up staying on the mountain behind the hospital. They spent one worry-filled night on the mountain. Yen remembered that she was very hungry and she did not have any money, not even a single baht. An ice-cream vendor came to where they stayed and he offered her ice cream and sticky rice without payment. After eating she felt much better.

They went back to the village the day after the tsunami to look for their house and see what was destroyed. Yen said that she was worried about the boat and when she found it in the mangrove forest it was completely wrecked. The house was also demolished and she said that not even a pole was left standing. When they came to the village, they saw people who had came to plunder the houses and steal what they could find. Yen said that there were dead bodies everywhere in the village but the thieves did not care, they took whatever they wanted. Yen found the bad smell from all the corpses unbearable and around their house there were about 100 dead bodies. The water had pushed them to their area and the corpses were without clothes. When Yen found what was left of their boat, she rescued the instruments and said they were lucky that nobody had stolen the costly equipment. Daeng and Yen stayed in the village for only a couple of hours. When they discovered that they had lost everything they had, they went back to the district office and were given some clothes to wear and some money. They lived in a temporary shelter for almost one year. Daeng said he felt completely empty inside and Yen said that she felt like crying all the time.

Neither of them had any relatives in the fishing village, but they did not consider leaving the place. They said that the fishing village was their home and they wanted to continue deep-sea fishing; they felt that this would not be possible in Petchaburi. Many of their previous neighbours died in the tsunami, and new people have now moved into the village. Their new place is not very far from their old one and they have a new boat and have continued fishing.

Pat

Pat lived with her family and her relatives in Bang Niang, close to the hill and far away from the popular beaches.[8] She was at home on the day the tsunami hit. She said that she did not know what had happened, but she saw many people and motorbikes passing her house. One of her cousins came by and said that they had to run to the hill because the seawater would flood the area, so she and her relatives headed to the hill. She described how she took a small bag with money and an ID card, and that was all. She went on a motorbike and she also took her sister-in-law to the hill. Her brother-in-law was watering the garden located in an area at the foot of the hill. He was not aware of what had happened and she told him about the disastrous waves from the sea. Then he told her what he had seen. He said it was amazing that, during a short time, the water in the canal withdrew and the canal was dry and then suddenly the canal was full again. He followed the rest of them to the hill and on the way they met a Burmese man and told him that the area was going to be flooded and asked him to run, but he did not. Pat said she thought that the Burmese people were not afraid. She also described how some of them took many things from the affected area. Not far from her house was her cousin's fruit garden. After the tsunami, the garden was crowded with many things, such as dead bodies and injured people, and also mud that had been pushed there by the water. Her cousin went back from the safe area to check her house and the fruit garden. She found many Burmese people looking for valuables. She found it strange that they did not assist the injured people even though they asked them for help. She met many who survived the tsunami and when they reached the hill they cried and cried and said that there was nothing left, all the houses were gone.

Her cousin was at home, washing clothes for the resorts and drying them in her garden. When they left for the hill there were a lot of clothes in her garden. The people who were hit by the tsunami but survived were often stripped naked and ran without any clothes on. She hoped they had used the clothes from her cousin's garden, because there were none left when they came back.

Pat mentioned one of her friends who owned a mini shop in Pakarang. Her friend and her husband left that area by car and they let a lot of people travel with them in the car. All the time in the car her friend called for the Buddha to help her. Before the tsunami, her friend had a very old car and she had bought a new one just before the tsunami hit. Pat said if she had tried to escape in the old car, she and the other people who went with her would not have survived. Her friend did not bring anything with her – she just jumped into the car and drove as fast as she could. Pat said that if she had worried about things like money and possessions, she would not still be alive.

Another of Pat's friends had a business, together with her husband, buying and selling squid. Unfortunately, they did not survive. Before the tsunami, the couple worried about their son who was at sea and the son sadly died in the tsunami. They had one grandson who was playing on the beach when the tsunami struck. The friend's husband and grandson were taken by the waves out to sea. The grandson said that both of them were holding on to a small log and his grandfather told his grandson not to let go of the log. The log was too small to support both of them. The grandfather died and they found his body after a year, but the grandson survived. He was floating at sea for two days and his back was badly burned by the sun.

Pat stayed on the hill for six days. She said that they could not be at home because there was no electricity and there was a very bad smell. Whilst on the hill, they slept on the ground. There were many foreigners who had also taken refuge on the hill and she said that they were extremely scared and many of them were injured. They did not dare to go down the hill to go by car to the hospital. Therefore, helicopters picked them up on the hill and brought them to the hospital. Pat was puzzled about why the well-educated foreigners did not know about the tsunami phenomenon. She said they seemed as ignorant as the locals were with only four years of schooling and they did not try to escape when the strange low tide occurred.

Manit and Ooy

Manit and Ooy, a married couple with two daughters, lived in one of the popular tourist areas north of Khao Lak.[9] Manit said that the day of the tsunami began as an ordinary day. He opened their shop and moved their car to the back of his house. When he was outdoors he met people employed at the hotels on their way to work. At about 10 a.m. he went outside again and noticed the very low tide. A friend called him and talked about how strange the sea was behaving and said that he had never seen anything like that. Manit called his brother to hear if he had any information about what was happening, but he did not know anything. Ooy's mother was in Phuket for medical treatment and she called her daughter and told her that there had been an earthquake in Indonesia. However, they did not think that the earthquake in Indonesia was related to the low tide in Phang Nga.

Ooy said that she thought the tsunami wave looked beautiful. To begin with, it was small and then it gradually grew in size. She described how many villagers came to the seafront to watch the spectacular wave. Some also caught the fish that were lying on the dry seabed when the water had withdrawn more than 2 km. Ooy said that at first she did not think it was dangerous and she did not consider leaving the place. She said she drove the car in order to avoid the sunshine. She was in the car when she looked over to the cape and saw cars racing at high speed, people running and a lot of water. She went home and let her husband, their two children and some villagers get into the car and Manit drove as fast as he could to the hill and did not stop until he could not drive any longer. Their daughters sat in the backseat and they saw the car driving behind them being pushed by water; they also saw a child separated from its mother by the water. The mother survived but the child perished. Manit said that he could never believe that a tsunami could be that dangerous. He mentioned that he said to one of his daughters when she tried to close the door to the house before they left that she did not have to bother because they would be back soon.

Manit and the family arrived safely at the hill and they slept outdoors under the trees. They were provided with food and water from his younger sister's shop on the main road. In the afternoon, some villagers who had survived the tsunami at sea came to the hill looking for their families. They were the first who had seen the consequences of the waves. They said that everything was in ruins and they thought that everyone had died. A friend told Manit that only the floor remained of his house. Manit could not believe that, but his friend said that Manit's own house was also demolished. On the second day of living on the hill, all of them, especially children and old people, were exhausted. Manit decided to look for somewhere to stay close to the main road. He went with a group in his car, but there was a rumour that a new tsunami was on its way and therefore they went back to the hill again.

The villagers had already begun to collect dead bodies on the same day that the tsunami struck. Manit said that family members are considered to be the most important and even if people were afraid, they tried to find them. On the hill, the villagers cooperated and helped each other in all the ways they could. They shared the food that they had and it was difficult to live there especially since the weather was cold and they lacked food and clothing. He remembered a foreigner who did not have any clothes and they modified a rice sack for him to wear. Manit said that on the third day the dead bodies were moved from the location where their house had been. Donations had also arrived and donated materials were placed at the police station. Manit said that he saw many people who were not affected by the tsunami taking what was donated. In that area, most people who were affected did not know about the donations because they had moved to live on the hill. They did not have any means of transportation and they did not have any money. Manit said he had 400 baht in his pocket and that was all.

The family lived on the hill for three days and then they moved to live at a children's centre. Manit described how he went back to the hill and invited some Moken people to stay at the children's centre. Manit and Ooy had a close relationship with the temple not far from where they lived. However, the temple was very near the tsunami-affected area and it was not safe to stay there. Most of Manit and Ooy's neighbours lost family members in the tsunami. Many lost their young children and they suffered greatly, but they refrained from talking about their loss. They wanted to forget and Manit said that some began to drink and others could not sleep and wandered around at night.

At temples

Amara

On 26 December 2004, Amara visited Wat Nangyon together with her brother and his wife.[10] The temple was a small *samnak song* located in Kura Buri district and Amara used to visit it on *wan phra.* They had just finished the morning chanting when Amara's sister-in-law received a phone call from her sister who told her about a giant wave in Tablamu. That was difficult to believe because the weather was so nice. Her brother and his wife left the temple at once and were on their way to Tablamu but when they found out that their son was on the island of Koh Phra Thong, they stopped at Green View Resort and went back to Kura Buri. They tried to get in touch with their son on the phone, but it was impossible.

Amara remained at the temple and people received messages about the tsunami. One of Amara's relatives and neighbours got a phone call from her son who worked in Khao Lak and who told her about the tsunami and that he had survived because he was on the second floor of the building; however, his wife, who was downstairs, had died. Amara asked her brother to take her neighbour to Khao Lak, but the road was closed so they could not go. They went to Kura Buri pier to wait for Amara's nephew. Koh Phra Thong was badly hit and those who survived the tsunami were brought to the mainland. Amara said that they were crying, they had fought for their lives and family members had died right in front of them. Most of the survivors went to the temple, Wat Pasan, whilst some went to the police station and some stayed at the pier. She said she felt so sorry for them.

The survivors talked about the water level that changed drastically from very low to very high. Her brother stayed at the pier until it was night and she said that her sister-in-law wept constantly. Rescue cars with sirens were heard the whole day and evening. They went back and forth between the affected areas, hospitals, temples and rescue centres.

Amara had a cousin who lived at Koh Phra Thong with his wife and a two-year-old son. The parents survived but their son died in the tsunami. The cousin and his wife were evacuated to the mainland and her cousin went back the next day to search for their son. He found his son's body and he said that the search was not that difficult because the body was covered by sand, and almost at the place where they were hit by the tsunami. Her cousin was afraid of going back to live on the island and therefore he stayed on the mainland. He and his wife divorced after the tsunami and later his cousin married a woman who had lost her husband in the tsunami and who had a child.

Amara's nephew, however, was difficult to find. The parents went to a temple and they conducted ceremonies with the hope that he would be found. They said that even if it was only his bones that were left it was important to find his remains. For three months, the parents went to the island almost every day to search for their son's dead body. They also consulted a medium that they brought to the island. To their relief, after three months, they found their son's remains.

On the morning of the day the tsunami struck, Amara's sister was at Wat Pasan to make religious merit (*tham bun*). When the ceremony was over, some villagers were eating and some were talking with the monks. One of the villagers had a phone call from her son who said that the Salasin Bridge between Phuket and the mainland was destroyed. She told everybody the news and they thought it was strange. Shortly after that message, people from Koh Phra Thong begun to arrive at the temple. The villagers from the island said that they did not have a house to stay in any longer and that Koh Phra Thong was totally destroyed. They came to the temple without bringing anything with them and they wanted to stay at Wat Pasan. The monks asked the villagers from the mainland not to go home but to stay at the temple and cook for the survivors from Koh Phra Thong.

Amara's sister was worried about her nephew, who worked at Koh Phra Thong, and she tried to find out what had happened to him. There was no news about him and she decided to stay at Wat Pasan. After that she went to Wat Pasan every day to help out with different practical matters that had to be solved. She said that she initially had difficulties seeing all the corpses that were brought there. The cremations of the bodies were not carried out in the crematorium but were performed in the open air because of the large number of corpses.

A temple on Phang Nga's mainland, in one of the worst-hit villages, was located behind a large pond and the wave did not reach the temple grounds. At the time of the tsunami, they had four resident monks but no *mae chi*s living at the temple. The abbot said that after the tsunami, people came running to the temple and they had about 1,000 people staying there. The waves had taken their clothes and at first the temple had to provide people with both clothes and rice to eat. Some people went to stay in other areas further away from the sea, for example, in the national park up in the mountains of Khao Sok, and they returned to the temple when the temporary shelters were set up. The temple had temporary shelters for 80 families. Most of the families stayed for six months and about ten families stayed for more than a year. The families who stayed longer did not dare to return to their houses. They were scared and some of them said that they could not afford to live outside the temple, since they had lost their livelihoods.[11]

Phuket

Luang Pho was 57 years old and he lived at a temple close to the sea in Phuket province. He ordained late in life and had been a monk for five years when the tsunami hit.[12] He described how the water rose and was full of bubbles. The whole building, including the monks' rooms (*kuti*s), was destroyed. He said that he was in his *kuti* when the water rushed in and filled the whole room up to the ceiling. The water came with a tremendous force and made a roaring sound.

The monk said that the first tsunami wave entered the temple's *sala* at the same moment that three old laywomen in their eighties were about to offer food (*dana*) to the monks. He explained that they had not gone on an alms round (*binthabat*) that day because it was *wan phra*. The monk said that he jumped over the laypeople in order to escape the water. The wave hit the laypeople and the old women ended up in the canal beside the temple. The monk used his outer robe and threw it into the water. The women held on to one end of the robe and the monk on to the other end, and he was able to rescue them by pulling them up on to the land.

After the monk had rescued the laywomen, he ran to the building housing the *kuti*s and when he entered his *kuti*, the tsunami struck a second time. He said that on this occasion the wave was much more powerful. He estimated that there was about five minutes between the first and second wave. The water rose rapidly to the second floor of the building and the wave destroyed the house completely. When he was in the water he recalled the Buddha, Dhamma and Sangha and then he recalled his mother. The monk explained that his mother was his first Buddha. He said that he wondered what bad *kam* he had done to experience this. He was fighting in the water and chanted "Buddhaguna Itipiso" and continued calling for his mother to help him. He could hear the sound of the wave smashing the cars that were parked outside the house.

Luang Pho described how, when the tsunami wave struck for the third time, he was pulled under the water. He found an electric wire and held on to that, unable to escape. His robe had disappeared and he was wearing only the sleeveless one-shouldered waistcoat (*ungsa*). He said that at that moment, something strange happened; a novice monk's robe (*samaneri*'s robe), which he could use, came floating from the storage room. The monk said that the third time the wave hit, he was totally exhausted but luckily the wave was not so powerful.

There used to be seven monks at the temple. On that day, two young monks were in Phuket town teaching and five monks where at the temple. Three of the monks in the temple died in the tsunami. The monk I talked with was injured and went to the hospital. At the hospital he had a very high fever and difficulty breathing. According to him, 69 people died in the village. Eighteen were Thai and 59 were from other countries; many more were missing.

The monk had never heard of a tsunami before it hit his temple. He decided to learn about the phenomenon and read a book about earthquakes and tsunamis and he said that he now had a strategy if a tsunami should happen again; he should climb the tamarind tree close to the *sala.*

Temples on islands

At a small temple on an island in Phang Nga, the head monk said that he sat on the floor in the *sala* when he heard a tremendously loud sound and thought that it was a fire in the forest. When the wave reached the temple it was 6 m high. He said: "At 10.45 a.m. the water was very high and at 10.55 a.m. very low. High five minutes and low five minutes, and it cost us two million baht".[13] The villagers had come to the temple to donate food and most of them survived. However, six laypeople died. The temple was so badly damaged that it was impossible to stay there and they could not have any temporary shelter at the temple site. All people on the island were evacuated to the mainland and when they later returned to the island, laypeople and monks worked together and restored the temple.[14]

A monk at another small temple (*samnak song)* on the same island said that he was eating when he heard that a tsunami was going to strike.[15] It was the village leader who came to the temple to evacuate the monks and the villagers who had gathered in the *sala* to the mountain. The village leader told them that they were expecting a tsunami and they did not know what that was, and the monk said that he had wanted to go to the beach and see what would happen, but the village leader stopped him and said that it was too dangerous to go to the beach. The monks went to the hill and stayed there until the evening. The temple was not greatly damaged by the waves, so he and another monk went back and slept in the temple the night after the disaster.

The day after the tsunami, the village leader reported that many people were missing. The monk went to the villages close to the sea and walked around for one-and-a-half hours without meeting a single person. He saw how damaged the village was, but he did not see any dead bodies. The following morning some villagers came and told him that everybody must leave the island. He had not seen any boats since the tsunami, but the villagers said that they should not leave by boat. All of them were going to be evacuated by helicopter, including the dead bodies. On each trip the helicopter took seven or eight people on board, beginning with the people who were injured. There were many rumours about a new tsunami, and those who were not yet evacuated wanted to stay on the mountain. The monk said that he would have preferred to stay at the temple but the villagers did not dare to follow him, so he had to stay with them on the mountain.

Twenty of the houses on the beach were destroyed and the monk told the villagers to go back and check their houses. Some villagers went and to their dismay some of them found that their houses had been looted.[16]

A whole village at another of Phang Nga's small islands was wiped out by the tsunami. The villagers who survived were rescued and taken to the mainland.

The village had a small temple with three monks, of whom two were killed. One month before the tsunami hit, the monk who survived the disaster dreamt that the principal Buddha image at their island temple had blood on its face. He said it meant that something bad would happen. The monk was in Phuket to collect a donated Buddha statue when the tsunami waves struck the island and when he came back nothing was left in the village. The villagers had been looking for the two missing monks but they could not find them. The monk who returned to the island found his fellow monks' dead bodies almost immediately. He said that they were not under sand; they lay beneath a tree not far from the temple.

The monk who returned to the island was 32 years old when the tsunami happened and he had been a monk for 11 years.[17] He was born in Phuket province and said that he was studying at the university when he decided to leave the lay life and become a Buddhist monk. He received ordination in the north of Thailand and lived there for seven years before he chose to return to the south, where he had lived for two years in the small *samnak song.* A villager had donated land with the purpose of establishing a temple on the island, which had long been a wish of the community. Before 2002, there was no Buddhist temple on the island. The school did not have enough teachers and the monk who had studied at the university helped out by teaching mathematics and other subjects at the school.

The monk went on an alms round every morning and he came to know the people in the village quite well. He organized temporary ordination for the villagers on their request because they wanted to make merit. The monk said that some of them were anxious because they were fishermen; they knew that it was against the Buddhist precepts to kill and that killing created *bap* (demerit). The monk supported the community and he said that he helped the villagers to develop their minds. He explained that the villagers were born Buddhists but have not had a close relationship with a temple. The monk remembered that when he first came to the island, the villagers were very interested to talk with him and they sought explanations from him as well as participation in Buddhist ceremonies and activities at the temple. After learning more about Buddhism, some villagers ceased to fish on *wan phra.*

After the tsunami, the monk worked side-by-side with the survivors and volunteers to collect dead bodies and transport them to Wat Pasan on the mainland. The monk said that he applied the Buddhist way of being in the present moment (being mindful) and he emphasized the importance of encouraging the villagers and trying to make them happy.

Concluding remarks

To sum up, all the people that I talked with in the affected areas said that the tsunami happened unexpectedly. Nobody was prepared for any hardship at sea or difficult weather on land. The spontaneous comment from the survivors was that it was a bright morning; there were no clouds in the sky and no wind at all, and some said they went out fishing because of the very good weather conditions.

In this chapter, I have chosen to relate narratives that display a variety of the men's and women's experiences of surviving the waves. In interviews and informal talks with those who were affected by the waves, they mentioned that they made crucial decisions almost instinctively and instantly. They pinpointed one or two examples of such decisions as vital for their survival. However, there are also examples of the opposite: survivors who recounted decisions that jeopardized their own and others' survival and who stated it was a miracle that they had survived. The survivors in this study were Buddhists and most of them explained that they survived because of their karmic status. By this they meant that their time to pass away had not yet arrived and they still had merit (*bun*) to live.

One female informant explained that people survived because they had *bun* that they had collected over many lifetimes. She continued by saying that they do unmeritorious deeds (*bap*) through their occupation (as fishermen) and that is because they do not know what else to do for a living. But when they have a break from their work at sea, they visit the temple and make merit, *tham bun*, to balance the *bap.* She said that they make lots of merit when they have a large income and they make merit on a smaller scale when their income is low. This is the mindset of thinking that she and many in the fishing villages live by and the economic situation is one sign of people's karmic standing. She said that they spend their income in order to sustain their lives now and for the future and this includes merit making. She continued, explaining about the balance between the bad deeds they do and the merits they earn through good deeds. She said that they survive as long as the weight of *bun* is heavier than the weight of *bap*. She emphasized that every person must take the consequences of all the *kam* he/she has created, both bad and good. Those who died in the tsunami, had finished their *bun.*[18]

Notes

1 Interviews with Wimon and Noi in 2006 and 2007.
2 Mina's narrative is compiled from interviews and conversations between 2006 and 2008.
3 Interviews and conversations, September 2006.
4 Conversations and interviews, May 2007.
5 Interviews and conversations between September 2006 and January 2007.
6 Interviews, September 2006.
7 Inerviews, September 2006.
8 Interviews, April 2007.
9 Interviews and conversations between December 2006 and February 2007.
10 Conversations in October 2006.
11 Interview, April 2006.
12 Interviews, January and March 2005.
13 Interview, April 2006.
14 Interview, April 2006.
15 Interviews, February 2007.
16 Interview, 28 February 2007.
17 Interviews in April 2006 and August 2006.
18 Interview, 22 September 2006.

4 A place of refuge

Organizing chaos

After the tsunami, the Buddhist temples became important places of refuge for thousands of survivors; they served as hubs around which everything functioned. This chapter describes the activities and work that was carried out at the temples after the tsunami and deals especially with the first month after the catastrophe. During this time, the temples operated as the main sites for the identification of dead bodies, the workplace for forensic specialists and sanctuary for many thousands of survivors. During that time, temples also provided a place to stay for many of the volunteers who came to help. Moreover, the temples were centres for information about missing persons and photographs of missing people were displayed with contact addresses for those who had information to give. This chapter begins with a brief outline of the daily routines and ordinary life at the temples and the relationship between the monks and the lay villagers. It ends with special focus on the Thai Buddhist nuns (*mae chi*s) in Phang Nga and their roles in the disaster situation.

Daily practices at the temples

The reciprocal relationship between the Buddhist monks and the laypeople is crucial. The Buddhist monks belong to the *lokuttara*, the transcendental realm, and the laypeople belong to the *lokiya*, the worldly and mundane realm. The *lokuttara* person is dependent on the members of the *lokiya* realm for material support and the laypeople are dependent on the ordained community for creating and transferring religious merits, during important passages in life and in times of difficulty. There were very few nuns in the temples in the tsunami-hit areas. As already mentioned, in Thailand the ordained state has been highly gendered and Thai Buddhist nuns, *mae chi*s, have never been included in the Thai *sangha.* During the recent decades, however, cooperation between Buddhist women internationally and *mae chi*s' improved educational standards have brought some changes for Buddhist nuns (Falk 2007).

Generally, Buddhist ordained communities in Thailand follow a similar daily schedule. The fixed routines are part of, and a basis for, the ordained life. Monks and *mae chi*s wake up early and their day begins at about 4 a.m. in the morning with chanting and meditation. The alms round is also performed in the early

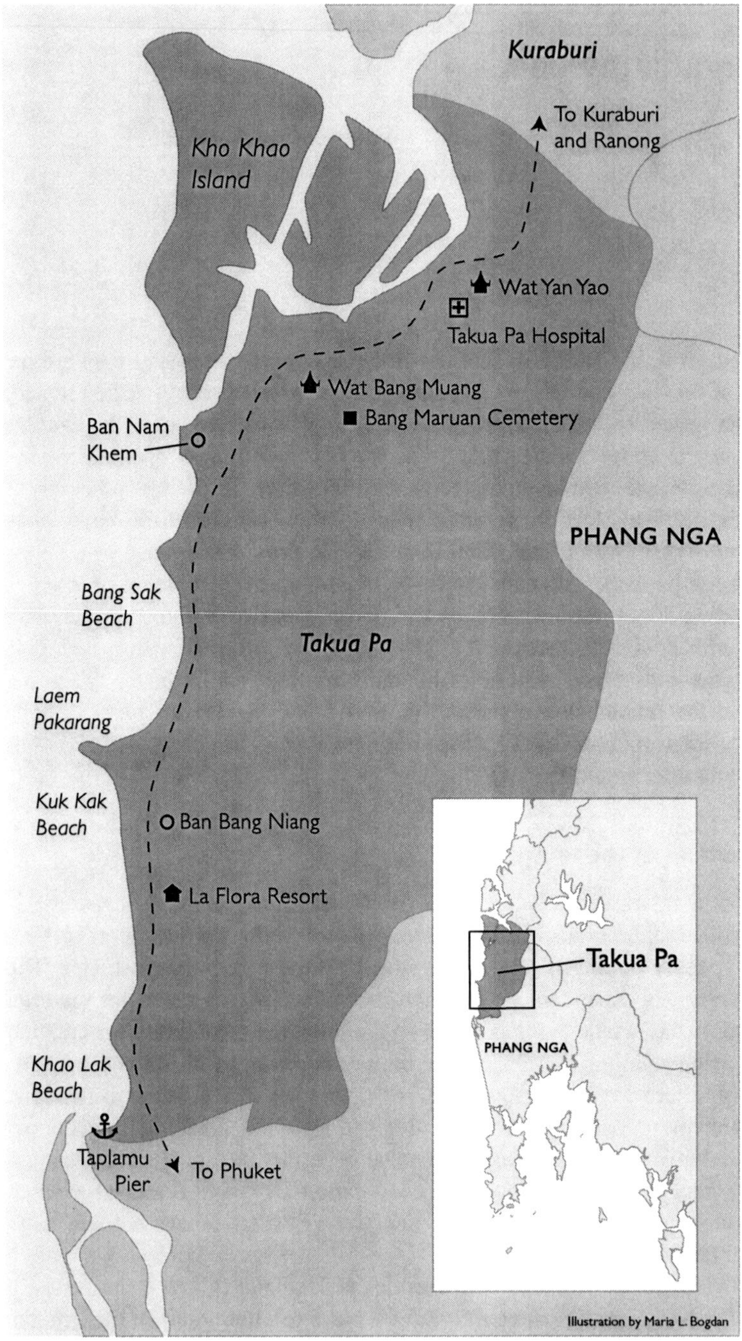

Figure 4.1 Map of Takua Pa temples (source: illustration by Maria L. Bogdan).

morning. The ordained communities eat once or twice a day. Both meals are before noon and then they fast until sunrise the next morning.

A minimum requirement is that monks and *mae chi*s pass the three basic Buddhist examinations (*nak tham tree, nak tham tho* and *nak tham ek*) (Falk 2007: 198). It is common that monks study Buddhist philosophy and Pali language at the Buddhist universities in Thailand. The Buddhist universities are now also open to *mae chi*s and laypeople. Monks and some *mae chi*s are also teachers and they tutor laypeople, young monks, other *mae chi*s and children. Monks conduct ceremonies at temples, but they are also invited to hold ceremonies at formal gatherings outside the temple and also in private homes. *Mae chi*s at *samnak chi*s who possess Buddhist education and skills in Buddhist chanting, would also be invited to monks' temples and to events outside the *samnak chi*s. Both monks and *mae chi*s provide services to laypeople, such as counselling and teaching.

Buddhist temples in Phang Nga

Some temples were destroyed by the tsunami and could not therefore be used. Temples that were still intact also provided a place to stay for many of the volunteers who came to help. Some temples gave shelter to people for a couple of days and up to a week. They stayed in the temples' *sala*s and thereafter many went to live with relatives, moved to temporary shelters or returned home. Other temples had people staying for more than a year. Several temporary shelters were put up in close connection with the temples. Most people stayed in the temporary shelters for about six months. There were villages on islands that were wiped out by the tsunami and the people who survived were advised to stay on the mainland and thus had to cope with totally new living conditions.

In Thailand, Buddhist monastic institutions are divided based on their official status. A *wat* is a temple with an ordination hall and/or a shrine hall. A *thipak* or *samnak song* is an unofficial/unsanctified "temple" or monastic centre lacking an ordination hall. *Samnak/thipak song*s commonly consist of a few *kuti*s (monks' and nuns' living quarters; often a small basic one-room house) and a *sala* (a multi-purpose hall).

In Phang Nga, there were 78 *wat*s and approximately 32 *thipak song*s *or samnak song*s. According to the national Buddhist register, there were 716 monks and 44 novices in Phang Nga. The temples[1] in Phang Nga had comparatively few Buddhist "nuns", *mae chi*s, and they were not documented in the records. In interviews with monks at several temples they mentioned that women were interested in ordaining but the small village temples did not have the required separate *mae chi*s' section and it was therefore not suitable for the temples to welcome *mae chi*s to live there.

Soon after the tsunami, surveys were carried out to catalogue the damage in the affected areas. One survey was conducted between 27 December 2004 and 5 January 2005; it investigated the destruction caused to the temples and *thipak song*s. The results were reported to the *Mahatherasamakom* on 11 January 2005. Studies were done to investigate the situation at the temples and to assess what the affected monks, novices and laypeople needed. The Office of National Buddhism conducted a survey about the effects of the tsunami in order to provide adequate support to the temples in the affected regions; it also created a plan of recovery for the temples and Buddhist institutions. The survey assessed the damage to property and identified which buildings had to be repaired or replaced. The team that conducted the survey also arranged meetings with the local monks, encouraged them in their work and made plans together with them for the recovery phase. They established a committee with the purpose of handling the money allocated for the damaged temples, Buddhist institutions, crematoriums and compensation for monks who had perished in the tsunami. Based on the preliminary results, about 50 million baht was mentioned to cover the costs.

The survey showed that no monks had died in Ranong, Trang, Satun and Krabi provinces, and the temples in these areas were not damaged by the tsunami. In Phuket, three monks died and two monks were injured; several temple buildings were demolished in Kathu and Patong districts. In Phang Nga, two monks perished but there were no reports on how many monks and nuns were injured. The survey listed that the following temples were totally or partially destroyed: Thipaksong Chareontham, Ban Pakjok, Tambon Koh Phra Thong, Kura Buri, Thipaksong Tungwha, Tambon Kuk Kak, Takua Pa, Samnak Song Nokna, Tambon Koh Kho Khao, Takua Pa, Thipaksong Paphothiwad, Tambon Kuk Kak, Takua Pa, Thungtuk, Tambon Koh Kho Khao, Takua Pa, Thipaksong Thungtuk, Tambon Koh Kho Khao, Takua Pa, Wat Panusnikom, Tambon Kuk Kak, Takua Pa, Samnak Song Thathai, Tambon Nateay, Thai Muang, Thipaksong Thathai, Tambon Nathai, Thai Muang, Wat Lakkhan, Tambon Lamkhan, Thai Muang.[2]

The same survey, carried out in mid January 2005, showed that the following temples handled dead bodies for identification and provided support to survivors and hosted volunteers in the temples:

- Wat Komneyaket, Tambon Kuk Kak, Takua Pa, received 100 dead bodies for identification and provided accommodation for survivors and 300 volunteers.
- Wat Rajniramit (also called Wat Bang Muang), Tambon Bang Muang, Takua Pa, received more than 1,400 dead bodies for identification and provided accommodation for survivors and 1,000 volunteers.
- Wat Nikornwararam (also called Wat Yan Yao), Tambon Yan Yao, Takua Pa, received more than 1,900 dead bodies for identification and provided accommodation for survivors and 1,500 volunteers.
- Wat Khongkapimuk, Tambon Takua Pa, Takua Pa, received 50 dead bodies for identification and provided accommodation for survivors and 100 volunteers.
- Wat Prathad Kiriket, Tambon Takua Pa, Takua Pa, received 30 dead bodies for identification and provided accommodation for survivors and 300 volunteers.
- Wat Samakhitham (also called Wat Pasan) Tambon Kura, Kura Buri,

received 100 dead bodies for identification and provided accommodation for survivors and 1,000 volunteers.

- Wat Ban Tream, Tambon Kura, Kura Buri, provided accommodation for survivors and 200 volunteers.
- Thipaksong Bang Krang, Tambon Bangwan, Kura Buri, provided accommodation for survivors and 200 volunteers.
- Wat Lakkhan, Tambon Lamkhan, Thai Muang, received dead bodies for identification and cremated 900 bodies after identification, provided accommodation for survivors and about 1,000 volunteers.
- Wat Prachumyothee, Tambon Thychang, Muang, provided accommodation for survivors and 100 volunteers.
- Wat Kittisangkaram, Muang, Phuket province provided accommodation for 3,000 people. Samnak Song Naithan, Muang, Phuket province provided food and essential materials for the survivors.

After the tsunami, the monks at the temples in the affected areas had to meet a tremendous demand for their services, not only in their capacities as religious specialists but also as leaders and organizers. The task of organizing all the activities that occurred at the temples had been hitherto completely unforseen by the monks.

The identification of the large numbers of bodies was a very difficult task and was initially carried out by Thai forensic specialists with help from volunteers. The number of dead people from more than 40 countries prompted the Thai authorities to cooperate with international forensic specialists. In Phang Nga, the first main identification site was at Wat Yan Yao and was led by the well-known Thai forensic pathologist, Dr Khunying Porntip Rojanasunan. When the corpses were identified, the monks held chanting for the *bangsukul* robe ritual and cremation ceremonies. At many temples in the affected area, the crematoriums cracked as a consequence of the constant heat. At some temples, they continued to cremate the dead bodies in the open air by using rubber tyres and at other temples they used wood, which is the traditional way of cremating dead bodies in Thailand.[3]

Identification and handling dead bodies

The large number of corpses was scattered in the vast territory of mainland Phang Nga province, which included inhabited as well as numerous solitary islands. Identification was performed in coordination with, and under the supervision of, the Institute of Forensic Science at the Ministry of Justice (Phang Nga: 1st Year Tsunami Memorial, 2005: 54). The government's forensic teams worked tirelessly alongside thousands of volunteers who had joined them from different parts of the country. The large number of dead necessitated international support and teams from more than 30 countries arrived to assist the forensic teams in Thailand.

The corpses were kept in three main mortuaries that were set up in the temple grounds at the Wat Lakkhan, Wat Yan Yao and Wat Bang Muang Buddhist temples, all located in Phang Nga province. These sites served as centres for the collection, separation and identification of corpses by the Institute of Forensic Science, assisted by Thai medical doctors and with international police officers standing by at the monastery of Wat Yan Yao. A team of medical doctors from a hospital in Bangkok, the Faculty of Medical Science, Mahidol University, Khon Kaen University and Srinakharinwirot University were on hand at the monastery of Wat Bang Muang and there was also a team of medical doctors from the Thai police on 24-hour standby at the monastery of Wat Lakkhan.

The temple of Wat Yan Yao in Phang Nga was transformed into a massive makeshift morgue. The forensic specialist, Dr Khunying Porntip Rojanasunan, led the huge identification operation together with volunteers and NGOs (see Rojanasunan 2012). Another makeshift morgue was set up at Wat Bang Muang, also at Phang Nga, not very far from Wat Yan Yao. The government issued an appeal to individuals and organizations for help in retrieving victims' bodies. DNA samples, fingerprints and dental records were obtained from each corpse to enable identification, including records of the bodies' condition and descriptions, such as any accessories etc., so that relatives could check and claim them. A DNA search and matching centre was established to enable the matching of DNA from relatives with that collected from corpses. The Ministry of Information and Communication Technology also set up an official website to make DNA information on tsunami victims available in order to help identify them. Authorities even installed microchips in about 1,000 bodies pending claims by relatives in order to facilitate identification.

The Disaster Victim Identification (DVI) unit was composed of specialists in forensic medicine, chemistry, dental medical science and experts in conducting DNA processing. The procedure of personal identification required a certain amount of personal information to conform to international standards, including fingerprints, DNA matches, and dental as well as medical records. In cases where this procedure did not result in identifying a body, clothes and jewellery were photographed, though this was not deemed a foolproof method, as bodies bearing similar marks or jewellery might be mixed up (Weber 2005: 49).

From 21 February 2005, members of the identification team collected readable fingerprints from more than 2,000 corpses. From the fingerprints of people reported missing, only 21 per cent could be matched with the corpses. DNA matching achieved a high rate of success and was recognized as a trustworthy method of identification. Because of the heat, bodies decomposed rapidly, and DNA technology was therefore considered to be the most preferable method in subtropical and tropical climates. The use of dental records, especially dental X-ray images, was another method employed to identify the dead. Photographs of missing people were displayed on websites and billboards, and databases with photographs of deceased people and their belongings were also shown (Weber 2005: 48–50).

A separate morgue, named the Thai Tsunami Victim Identification (TTVI) unit was set up to store corpses that were decomposed to the extent that physical identification seemed unfeasible. Mai Khao Cemetery, located in Mai Khao sub-district, in the Thalang district of Phuket province, was chosen as the site for the collection of unidentified corpses retrieved in Phuket and Phang Nga provinces. Mai Khao Cemetery was maintained as the facility for further identification efforts where corpses yet to be identified were kept in special containers. This cemetery was chosen for the convenience of transporting foreigners' corpses to Phuket International Airport and corpses of victims from neighbouring Phang Nga province to their places of origin (Weber 2005: 50). Permission for the release of corpses was conditional upon their identification. To receive a corpse at Mai Khao Cemetery, one had to seek approval by a sub-committee in charge of issuing the death certificate (ibid.: 50).

The Thai Armed Forces built storage depots for coffins. They aided dentists and medics who helped to identify bodies and they assisted in sanitizing corpses at the morgues set up in the premises of the temples at Wat Yan Yao and Wat Bang Muang, and in the Bang Marang Graveyard in Phang Nga province.

Wat Yan Yao

Dr Porntip and her team arrived in Phuket on the same day that the tsunami struck. They thought that Phuket was the most severely affected province. Dr Porntip's team consisted of eight people, and five of them remained in Phuket when she, together with two assistants, went to Phang Nga to investigate the situation there. She writes that they were insufficiently equipped with just a laptop, two cameras – one digital and one film camera – and a notebook (Rojanasunan 2012: 168).

Wat Yan Yao, which is located in the town of Takua Pa, became a centre where dead bodies were brought for identification. When Dr Porntip and her team arrived, more than 500 corpses had been placed in the different temple buildings and also laid on the ground. Dead bodies were brought to the temple and piled up. The situation was chaotic and the work was not organized. The forensic team's main task was to make records for identification and hand over the bodies to their relatives as soon as possible. The bodies that came to the temples in the first couple of days were relatively intact but because of the hot climate, they started to decompose quickly. The team was working against the clock and tried to work as fast as possible. In Phang Nga most of the people who died were from other countries. The identification procedures for Asian people and Caucasians were carried out slightly differently (Rojanasunan 2012: 170). They photographed the bodies and for the Caucasians, they also took DNA samples. Later they also used DNA analysis for Asian people. DNA was collected from surviving family members.

The team faced huge problems in finding a way to mark the bodies so that they would be traceable. They were not provided with suitable tags, and therefore Dr Porntip first made labels with identification numbers and basic information written in indelible ink on paper. Strips of paper were then tied with twine to the wrist of each corpse and photographs were taken. A couple of days later the team found that the labels had absorbed fluids from the bodies and the information had become blurred. They tried to solve the problem by putting the labels in sealed plastic envelopes. However, after a week the fluids from the bodies were leaking into the envelopes. The chemicals that the corpses were sprayed with to protect them from insects made the labels deteriorate. They decided to make three labels for each dead body. One label was placed on the corpse, one on the body bag and one on the marker that was put up where the body was buried. They did not have any instructions from the authorities as to how they should handle the bodies. The forensic team decided to temporarily bury the bodies that they had photographed and documented in order to keep them from the sun and to provide space for more bodies that continued to arrive at Wat Yan Yao. They needed to find a better way to mark the bodies and so decided to use microchips. After consultation, they implanted the microchips in the sinus cavity. However, that proved to be too deep to be read by the scanner and they had to once again try to find another method. They ended up using metal plates that were fastened onto the bones (Rojanasunan 2012: 173–176).

While the forensic team worked with autopsies and struggled to find a satisfactory way to make the bodies traceable, the relatives and surviving family members came to the temple searching for those who had perished. Mina, who we met in the previous chapter, was searching for her mother, brother, youngest son and other relatives who were missing after the tsunami. As told in Chapter 3, she had been at sea with her husband when the tsunami struck. They survived the waves and managed to return to the village the same day and in the afternoon they found that their daughter and oldest son were safe and well. The village where Mina and her family lived was one of the worst hit in Thailand. Mina said that other families found their relatives' bodies but her family could not find their deceased. She went to Wat Yan Yao and also to other temples in the area in search for her family members and also for other people who she knew were missing and she said:

> I had looked at every dead body that was brought to the temples to see if it was my mother or my son or somebody else that I knew was missing. I did not mind the terrible smell and the yellow liquid. I opened the white cloth that covered each body trying to find my family members' bodies.[4]

Mina found a body that she thought was her mother's and she went together with her niece to Wat Yan Yao for identification because she was uncertain about the identity of the corpse. Most of the dead bodies were badly damaged and it was almost impossible to identify them by looking at the features. Mina could not remember the pattern of the skirt her mother wore that day and, therefore, she was unsure. She said:

> I marked that body with a piece of string around her wrist and later when it was possible to use computers with numerous pictures of each dead body I realized that it was my mother's body that I had found and that was also proved by DNA.[5]

When the system of identification and labelling of the bodies was in place, it became much easier for the relatives to search for missing persons. The forensic team was under-staffed and short of equipment. Volunteers used their own cameras and they also borrowed cameras from the villagers. However, the quality was not optimal since most of the cameras had resolutions of four or five megapixels and forensic work requires a resolution of at least eight megapixels. DNA played an important role in identifying the corpses (Rojanasunan 2012: 176). The authorities decided to send DNA samples for analysis to China and this meant that the forensic team had to exhume the bodies that were buried in order to take new samples, according to the requests from the laboratories there (ibid.: 178).

The police set up small offices at Wat Yan Yao and Wat Bang Muang for relatives who came to collect the bodies. Bodies were moved from Wat Yan Yao to Wat Bang Muang and cremations were carried out there from early morning to late at night.

After the initial difficulties in preserving the bodies in the humid climate, they were documented and buried. Shortly after that, Wat Yan Yao received refrigerators in which to store bodies. The refrigerators made it possible to preserve the bodies until they were returned to their relatives. However, it sometimes took a long time and some people had to wait for more than a year until they received their relatives' remains. The forensic team at Wat Yan Yao had worked tirelessly for about a month when the Thaksin administration decided that the police should take over the identification process. The forensic team had collected data and documented information on almost 4,000 bodies. At the end of January, they closed the identification centre at Wat Yan Yao. The bodies that were kept there were moved to Mai Khao Cemetery, located near Phuket Airport (Weber 2005: 50).

Wat Bang Muang

Some temples were crowded with people seeking refuge and other temples were overloaded with dead bodies. Wat Yan Yao and Wat Bang Muang were the two temples that had the largest amount of bodies and both temples worked in close cooperation. They cremated the bodies at Wat Bang Muang instead of Wat Yan Yao and many of the bodies that were brought to Wat Yan Yao for identification were later transferred to Wat Bang Muang.

A monk at Wat Bang Muang said that the situation at the temple was extremely inconvenient during the time after the tsunami. The monk was from Nakhorn Si Thammarat province and was ordained at Wat Bang Muang when he was 14 years old; he had been a monk for 17 years. At the temple, there were 12 resident monks and three *mae chi*s. The monk said that on the day of the tsunami, there were many laypeople at the temple merit making, *tham bun.* The monk heard some people shouting that they should evacuate and he thought that it might be a fire. The temple is located 3 km from the village where more than half of the villagers perished. People who lived closer to the sea came running to the temple to take refuge. The police said that the temple was not a safe place because the tsunami might reach there, so people moved to safer ground. The monk said that he had responsibility for the temple and could not leave. He had planned to take refuge higher up in one of the temple buildings if the tsunami reached the temple.[6]

There were about 1,500 dead bodies at the temple grounds. They buried the bodies because of the hot weather, which would make the bodies decompose quickly. Their health was badly affected and they found the smell from the corpses almost unbearable. The monk said that they could not move. He explained that they had to stay at the temple because they had to be with the dead bodies and receive all the people who came to visit. He said that from 27 December for the following three months, everything at the temple was dirty and useless. But after another three months, things became better and they received assistance from monks and lots of other people. Monks from other temples came and helped them with the cremations. He described how they started to cremate the dead at 6 a.m. in the morning and continued the whole day until 11 p.m. at night. He said that they had sore throats from all the chanting. They had difficulties in handling the large number of bodies. He added that other temples had additional difficulties: “We had only dead bodies here. Dead bodies do not eat and do not need all the things that people who live at the temples for a long time need”.[7]

The monks mentioned that people came to the temple and asked for help to go on with their lives; they sought something that could make them brave, encourage them and make them feel better. They appreciated receiving a Buddha image or sacred thread (*sai sin*). Laypeople asked the monks for explanations about the tsunami. The monk said that he clarified that the tsunami was the consequence of a natural phenomenon, an earthquake. He said: “We have done unwholesome deeds but the tsunami was natural. People are acting in ways that are harmful for nature, like nuclear tests and sometimes nature struck back”.[8]

Under normal circumstances, the monks at Wat Bang Muang woke up at about 4 a.m. every day. The monk I talked with said that they start the day with Buddhist chanting followed by sitting and walking meditation. The monks performed alms at 6 a.m. in the morning and this particular monk said he walked less than 1 km on his alms round. Back at the temple, he used to clean himself and have a short rest before breakfast. The monks at this temple had two meals a day. Their first meal was at 7 a.m. and they had the last meal at about 11 a.m. After breakfast, the monks cleaned and swept the area. The monk said that: “If the monks do not do that they will be reborn as a buffalo or cow to help the farmers”.[9] The temple had a school for novices during the rainy season, which lasts for three months, and the older monks gave the young monks Buddhist (*dhamma*) lectures. Every evening at 7 p.m. the monks and the *mae chi*s gathered for chanting followed by meditation, which is called *tham wat yen.* They went to sleep between 9 and 10 p.m.

After the tsunami, the monks could not keep up the daily routine of the alms round. The monk I interveiwed said that it was a time of confusion. The villagers had lost their properties and many had lost people who were close to them. The villagers who survived the tsunami had fled into the mountains because they were afraid that a new tsunami would occur. The shops were closed and the monk said that they ate instant noodles. They continued with the morning chanting and meditation but from 6 or 7 a.m. until midnight, the monks were conducting cremation ceremonies.[10]

The temple had one crematorium but that was not enough, so they also performed cremations in the open air as in the old tradition. At Wat Bang Muang, like many other temples where cremations were conducted throughout the day, the crematoriums cracked because of overheating. The temples later received new crematoriums in compensation for the old ones. The monk said that it was their duty to carry out the cremations and that it was important for the relatives and for the dead person.[11] At Wat Bang Muang the monks were divided into two teams and monks, novices and volunteers all worked together performing the cremations.

Several monks were critical of how the donations were handled. They said that only a small part of the donations went to those affected by the disaster. The monks at Wat Bang Muang said they saw people coming on motorbikes and in cars who stole donated goods right in front of their eyes. One monk commented that those people who acted selfishly and stole from the affected people were not afraid of going to hell. They even stole the special white fabric that was used to wrap the corpses.[12] Another comment was that if the coffins had been smaller so they would fit into a car or a bag, the thieves would have also stolen the coffins. The monks also felt pity when they saw the face of human defilement (*kilet*) that, according to Buddhist teaching, leads to suffering.[13]

Material donations were of vital importance for sustaining the lives of the survivors. Many had lost everything they had and they needed all the basic things, including a means to make their livelihood. The villagers also needed encouragement and help with their inner feelings and to find ways to cope mentally with what had happened to them. Some villagers went to Wat Bang Muang and asked for a talisman or a Buddha statue. One monk said that he had distributed many talismans and he also brought them with him to the temporary shelters and gave them to the survivors. The monk said that the talismans helped the survivors to recover their sanity.[14] The monks supported the survivors and stood by them in their suffering. In contrast, one monk concluded, the survivors had received no visits from community or village leaders from the eight villages in the district that included Wat Bang Muang, even though it was a centre for keeping the dead bodies after the tsunami.[15] They might have been too busy with other matters, but the villagers had told the monk that the leaders were afraid of infectious diseases. The monk praised Dr Porntip and her team and said that they worked in an exemplary way.[16]

The monk explained that in times of difficult, they are needed by villagers in order to reassure them. After the tsunami, villagers invited monks to avert "bad luck" by sprinkling holy water at their houses. The villagers also wanted to have the straws used for sprinkling water in front of their houses, because they believed that this could obstruct evil spirits from entering their homes and they also believed that the holy water could enable their family members who had perished in the tsunami to reach the heavenly realms.[17]

Mai Khao cemetery

One month had passed since the tsunami struck, when the authorities began to transfer corpses from the temples in Phang Nga to Mai Khao Cemetery in Phuket province. The area was an existing Moken burial ground which was renamed the International Repatriation Centre (IRC). Almost 2,000 bodies of foreigners were still unidentified and kept in refrigerated shipping containers. The forensic team that had worked in Wat Yan Yao had been dismissed and new teams of forensic specialists continued working at the new site identifying the bodies. This time, more staff were engaged to carry out the task and the working situation was much better than it had been in the chaotic first couple of weeks at the temples in Phang Nga.

The IRC site was opened on 25 January and Buddhist monks and representatives from other religions conducted a ceremony to commemorate the event and honour the victims. At the site, they had raised a wall named the Wall of Remembrance, which was blessed by the monks. The wall was white, without inscriptions and was used by those who had lost loved ones in the tsunami to leave photographs, flowers or other items to remember those who had perished.

Laypeople told me that most of those from Ban Nam Khem, Ban Bang Sak and Ban Bang Niang who died in the tsunami were kept at Wat Bang Muang. Many villagers were against moving the bodies from Wat Bang Muang to Phuket. The villagers had waited for identification of their missing relatives and family members and they found it unnecessary, inconvenient and costly to travel to Phuket.

At the end of January 2005, I visited Mai Khao and talked with a civil servant who reluctantly explained that they were going to take new DNA samples because they could not use the old ones. They planned to keep bodies from Phuket and Phang Nga at Mai Khao. On the day I visited Mai Khao in January 2005, 45 corpses laid in coffins had just arrived from Patong in Phuket. The coffins were placed on the ground at the site before examination. Later they were kept in refrigerators at Mai Khao. The civil servant said that the procedure for foreigners who came to Thailand to look for missing relatives was to first contact their embassies and after that they could visit Mai Khao for identification.

Supervising survivors and managing donations

Wat Pasan

There were five temples and five *samnak song*s in Kura Buri district in Phang Nga. The temples and *samnak song*s assisted the survivors and cremated those who had died in the tsunami. Wat Samakhitham, also called Wat Pasan, in Kura

Buri district, played a crucial role, especially for the Moken and Moklen survivors from Koh Phra Thong. Wat Pasan's abbot, Phra Kru, had long been a socially engaged monk and before the tsunami hit his temple, he had run social programmes. The monks took care of HIV-infected people and they ran an orphanage for boys whose parents had died from AIDS. Phra Kru had a Masters degree in Buddhism and Social Welfare from Maha Chulalongkorn Buddhist University and after the tsunami, he studied for a masters degree at NIDA, the National Institute Development Administration.

Wat Pasan is located about 7km from the sea. There were about 15 resident monks and the temple became a centre for many aid projects after the tsunami. A temporary shelter was set up on the temple grounds and I visited the temple frequently, finding the head monk's leadership skills to be extraordinary. This temple became a safe place and a source of guidance and hope for the future for many survivors, and I was amazed by the resident monks' ability to run so many projects and, at the same time, manage to carry out the usual activities.

About 100 dead bodies were brought to Wat Pasan and most of them were found on the island Koh Phra Thong. Phra Kru told me that after the catastrophe more than 1,000 people came to seek refuge at the temple. Both Thai people and foreigners came to the temple to look for their family members, relatives and friends that were lost.

The temple was approached by many aid organizations that wanted their aid programmes to run from there. That was a difficult undertaking for a small temple with few personnel resources and a large number of people in acute crisis that required support. The aid programmes, however, brought resources to the survivors and the temple agreed to take on the difficult task. The different NGOs had their own policies and target groups for receiving aid, and managing the projects required a lot of administrative work. The temple kept accounts of all the received donations and how they had been spent from the first day after the tsunami. There were rumours among the villagers that donations given to village leaders and temples got into wrong hands. Phra Kru was aware of the criticism and was particular about explaining to me that the temple's finances were taken care of in a correct way. He showed me the books the temple kept for the donations it had received and the books for the different projects' accounts. Phra Kru said that he had learnt that there were a lot of problems with corruption in the "aid-business".[18] The uneven distribution of aid and a lack of transparency in how aid was distributed were topics that were constantly brought up in discussions with villagers. They were very critical about injustice in the distribution of aid and they could give many examples of the corrupt use of donated money.

For about ten years prior to the catastrophe, the temple ran a shelter for up to 50 people with HIV/AIDS. The monk mentioned that many of them were prostitutes from Burma. He said that until anti-retroviral drugs were accessible, most of the infected people at the temple died. The HIV/AIDS shelter was located behind the temple. Those who sought refuge there did not have any other place to go; the temple was their last resort. After the tsunami, when all the survivors turned up at the temple, the infected people left the shelter. Phra Kru said that they felt that they could not stay there. Some went to the project Ban Phuan Chivit in Phuket province and Phra Kru considered it to be a good project, which was managed by one of his friends.[19]

The more than a thousand people who stayed at Wat Pasan needed food, shelter, medicine, clothes, etc. In the beginning, the survivors slept in tents which were borrowed from a local eco-tourist company that operated in the area. The Thai Red Cross and Hong Kong Red Cross donated tents that were used by the military. However, the tents were not a good solution because they could not be used when it rained due to leakage and those who stayed in the tents had to move into the *sala*. Also, the tents were too small for one family. During daytime it was unbearably hot inside the tents, and therefore most people sat in the shade under the trees; in the evenings there was no electricity and it was dark inside the tents. The temple tried to find alternatives to substitute for the tents. Another urgent difficulty that the temple had to solve was the sanitation issue, since the temple only had ten toilets.

Survivors from Koh Phra Thong

Some of the villages on Koh Phra Thong were hit hard by the tsunami; Village Four (Ban Pakjok) was completely destroyed. All houses, together with the school and the temple were demolished and people lost their livelihoods. Village Three (Ban Koh Ra) and Village One (Ban Thungdab) were less affected. Houses were destroyed, their livelihoods were affected and they lost their fishing boats, fishing gear and other utensils that were necessary for making a living on the island. Village Two (Ban Thapaeyoy) was the village that suffered least material damage. However, the villagers' livelihoods were affected and many had become too scared to live on the island.[20]

A temporary shelter was set up at Wat Pasan to help 32 families from Village Four. Most of these families were Moken and the families had 160 family members, 41 of which were children. Their village was completely wiped out by the tsunami. The survivors had lost everything and many of them suffered from Post-Traumatic-Stress Disorder (PTSD). A foundation (the Rak Thai) cooperated in building 25 temporary houses for them. These were stilt-houses made from plywood supported on concrete poles. The size of the houses was in proportion to the actual families that would live there. The monk told me that it did not take a long time to build the houses and the villagers were satisfied with them. Some of the families used four of the existing houses at the temple that were previously homes for the people with HIV/AIDS. The families from Village Three were on the government's list of affected people and they received support from the Thai government. The government contributed to the building of four toilets, a water-filtering system and a water supply system for the temporary shelters.[21]

The survivors needed something positive to do to occupy their minds with thoughts other than those of the catastrophe. The temple set up and organized different groups so that the villagers could come together and take part in useful activities. Among these were a sewing group, a group that made sweets, a boat-building group and a local boat model-making group. They earned some money from selling what they produced and participating in the groups helped them mentally. The group that made Moken boat models was also hired to build the temporary houses. The temple organized nursery care and schooling for the youngest children in the Child Centre at the temple and older children went to school in the village at the temple's expense.

The 55 families from Village Two went back to their village on Koh Phra Thong when it was considered safe to return. Village Two was not badly hit by the tsunami and these people still had their houses. The families, however, found it difficult to live on the island and they were afraid that a new tsunami would strike the island again. They had lost their boats in the disaster and thereby their livelihoods and there were also problems with authorization of the land. After the catastrophe there were constant rumours about a new tsunami and on 25 January 2005, the families moved back to Wat Pasan again. The temple had not prepared temporary houses for this group and so the families had to live in tents. The temple was criticized and accused of moving villagers to the temple with the purpose of attracting donations.[22] The families from Village Two did not receive much financial support. They lived in tents and they did their cooking outside, and there were problems with the outflow of water. The poor living conditions were stressful and that had an impact on their health. They received health care from Kura Buri hospital and personnel from the NGO ADRA's health office visited the shelter every two weeks.[23]

Individual survivors were searching for their missing family members on beaches and at temples where bodies were kept. Some villagers searched together in groups and there were also organizations and teams of volunteers looking for those who had perished. In contrast to Takua Pa, Kura Buri district carried out identification at the hospital and not at the temples, as was done especially at Wat Yan Yao.[24] Teams of volunteers organized by the district authority made coffins and distributed them to the temples. Monks, volunteers and relatives cleaned the bodies and kept them until the cremation service and the monks carried out funerals and cremations (funeral ceremonies are dealt with in Chapter 5). Monks, with help from volunteers, made surveys of the situation and kept records of the information they received. They also cooperated with many organizations. The preparation of food and distribution of donated material was carried out at many temples by women's groups. Medical doctors, nurses and public health officers administered first aid, medical and health care. Several temples arranged for survivors to live temporarily on their premises.[25]

Donations and distributions

People from the nearby areas contributed what they could. They gave clothes, food and also money. The temple handled donations from international non-governmental organizations (NGOs). This was not always an easy task. Phra Kru told about a Singapore–Hong Kong bank that had provided money for temporary shelters to be set up at the temple, but the money was blocked by influential politicians and authorities that did not want the donation to go to the temple. When a Thai royal princess paid a visit to the temple on her visit to the district she made some statements that were in favour of the temple and after that, the temple was able to access the money and construct temporary shelters.[26]

The temple had to put in a lot of effort to keep records and administer the paperwork. Reports were provided to all the donors. They set up contracts with donors and they also registered all donations. The people who received donations were also registered, with names, former addresses and notes about the size of the family involved being made. Every family that lived in the temporary shelter was given a book that they had to bring when they received rice, oil and other items that were distributed by the temple.

The temple also kept accounts on what every family had received. Life on the mainland was a new world to the Moken and Moklen people. They had limited experiences of using money and Phra Kru realized that they needed to learn how to handle money and how to organize their community. When the monks set up the shelter, they placed the weakest people, old women and men, in the centre of the community and the more capable around them. They learned to keep records of what kinds of food and other things they used every month and how much these cost. They learned to manage their lives by planning their food, expenses for clothes, expenses related to children's schools and saving money for themselves and for their families. Phra Kru said he advised them not to buy things that they could make themselves and it would be much better for them to grow their own vegetables. Every family that stayed at the temple received one boat and a saving account containing 1,000 baht. The money was a loan and they were supposed to pay back the money when they began to get income from their fishing. Phra Kru said that when they wanted to return the 1,000 baht he told them to save the money or use it for the family.[27]

Visits by Buddhist monks after the catastrophe

The *sangha*'s nationwide network proved to be invaluable in mobilizing monks from other parts of the country to travel to the south and help out in the temples. Initially, monks from 16 provinces came to assist the local monks at the temples in the affected areas. They helped out with cremations and ceremonies, and gave survivors support and psychological counselling. The *sangha*'s network was also effective in mobilizing financial and other resources to the impacted regions. The monks also arranged activities for the survivors as an aid to recovery. A rotation scheme was set up for monks who had worked around the clock with cremations and other duties and they could, if they wanted, have some rest.

Shortly after the tsunami hit, monks from the two religious universities – Mahamakut Buddhist University and Mahachulalongkorn Buddhist University – went to help their colleagues in the south of the country. During a couple of months, groups of university monks assisted the local temples and the survivors. The monks were divided into teams and then into smaller groups. The first team from Mahachulalongkorn Buddhist University went on 31 December 2004, with the mission of improving survivors' mental health. The monks had studied Buddhist psychology and mental health and their work was based on this learning. Monks and villagers worked together to improve the living conditions and livelihoods of the victims and the monks helped them to handle their grief and loss. The monks listened to the survivors, encouraged them, gave advice, and conducted ceremonies so they could transfer religious merit to the deceased, which was very important for the survivors. The monks had an advantage over the medical professionals, since many survivors were more open to talking with the Buddhist monks than with the psychiatrists and psychologists employed by the government who treated them as mental patients. A monk explained to me that if people go to see a psychologist they view themselves as mad, but when they meet a monk they look upon themselves as being meritorious, as they have *bun.*[28]

It was a difficult task to help out after the tsunami and the Buddhist universities selected monks based on the individual's knowledge, their ability to carry out the duty and their own ambition to help the affected people.

The monks were instructed by their university to be careful with food and drinking water so they did not contract any diseases. They brought medicine and other items they might need. If a monk fell sick, he would not be of any help to the villagers; on the contrary, he would be a burden and it would have been difficult for the villagers to take care of him. The teams were important for the groups of monks and they shared experiences and could broach the difficulties they had encountered. The monks gathered information about how the survivors felt, evaluated the results and followed up on the survivors' progress. The monks described themselves as "somebody who stands on the land helping people who are drowning". They emphasized that they themselves had to be without worries in order to be good friends (*kalyanamitta*), able to help and encourage the survivors.[29]

The fourth team from Mahachulalongkorn University were in Phang Nga between 19 January and 1 February 2005. They were divided into four sections and all of them worked on mental recovery. The monks visited all the temporary shelters and they also used radio broadcasts to communicate with people. A survey that the teams carried out in 16 temporary shelters showed that most of the survivors were recovering. There were severe cases that were transferred to the mental health centre for further treatment and other cases that needed medication and follow-up investigations. The monks concluded that the situation was not stable and the survivors' labile mental state in combination with having lost family members, relatives, friends, all their material stability and their livelihood, made their recovery slow. Many of the survivors did not have any idea how they could carry on with their lives and the monks said that these people needed regular support. Victims could see that ceremonies and consultations with the monks had a positive impact on their "mind recovery". The monks' expertise concerned the inner recovery, not the material.[30]

Asoke has centres all over Thailand, which cooperated and sent monks, ten precept-ordained women (*sikkhamat*s), and eight precept male and female lay followers to the disaster-hit provinces (see Chapter 9 on the Asoke group's activities after the tsunami). Asoke members travelled in buses and the relief action was well planned by a group of monks who had gone to examine what the survivors needed. They brought tents that they lived in during their mission in the south. In Phang Nga, they set up eight centres and helped the survivors with all kinds of practical matters, as well as with their mental recovery. In Takua Pa district, about 80 monks camped on the grounds of Phra Narai in Takua Pa town. They came to encourage those who had lost family members in the tsunami and help to decrease their suffering. Monks that I talked with said that they did not distribute any amulets or talismans but instead issued Buddhist books, which taught how people could solve difficulties with wisdom.[31]

Gender and Buddhism

In Thailand, the *sangha* is solely male and it forms the monastic community of Buddhist monks and novices. Ordained life is strictly regulated and creates a sharp boundary between the ordained community and the laypeople. There have always been *mae chi*s in Thailand, but they do not belong to the Thai *sangha.* In ancient times, when Buddhism was formed, the Buddha established a female ordination lineage five years after the first males were ordained. In Pali the male monks are named *bhikkhu* and the female monks are called *bhikkhuni*. The *bhikkhuni* lineage spread to Sri Lanka, but it is uncertain if the *bhikkhuni*s went to South-East Asia (Kawanami 2007). The Thai *sangha* has never ordained women. The *mae chi*s' exclusion from the *sangha* implies that they are eliminated from the benefits that their male colleagues enjoy. The *mae chi*s' position outside the *sangha* makes their status somewhat ambiguous. Their position is context-dependent on whether they are considered to belong to the religious or the lay realm. However, exclusion has given some *mae chi*s the opportunity to create a space for themselves and establish their own *samnak chi*, "temples" and thereby better fulfil their roles as ordained persons (Falk 2007). The *mae chi*s' lack of formal authority in the religious realm and their comparably weak organization, compared with the monks, hindered them from helping out in the aftermath of the tsunami in the same way as the monks were able to do.

The *mae chi*, and the *bhikkhuni*, are two categories of Buddhist monastic women that exist in Thailand today. *Bhikkhuni*s are novel in the present-day Thai religious realm. *Mae chi*s, on the other hand, could be seen as a substitute for a missing female ordained category and they have existed in Thailand for centuries.

In contrast to the monks' well coordinated aid contribution, most of *mae chi*s who went to help the survivors did so individually and in the company of laypeople. Women's exclusion from the *sangha* and the lack of full religious recognition affect gender inequality both in the religious field and in Thai society as a whole (Falk 2007, 2008). Compared to other regions in Thailand, there are few *mae chi*s in the areas that were hit by the tsunami. This meant that *mae chi*s in other parts of Thailand who wanted to help the survivors had nowhere to stay;

they could not stay at the temples because most temples in Phang Nga and Ranong provinces were lacking both *mae chi*s' sections and "nunneries". I observed that female survivors especially appreciated talking with the few *mae chi*s who visited the area. The *mae chi*s explained to me that the women felt a closeness to them that they did not feel with the monks because the *mae chi*s were women. Laywomen could talk about certain issues with the *mae chi*s that they did not bring up in conversation with the monks.[32]

Thai Buddhist "nuns" in the wake of the tsunami

Most of the *mae chi*s who went to help the survivors went individually or together with laypeople. Some *mae chi*s that I have interviewed at temples in Bangkok, and in provinces in Isan, expressed uncertainty about their ability to help. Their doubts focused mostly on practical matters, for example, where to stay, because there are very few *mae chi*s at the temples in the affected areas. One exception is the *samnak chi*, that is, the Thai *Mae Chi* Institute's branch in Phuket where 60 *mae chi*s live. In an interview with some of them, they said that they had initially considered going to the affected areas to help the survivors and were confident in their capability of doing so. However, after discussions with the rescue teams, they decided to concentrate on cooking food at their *samnak chi* for distribution to the survivors. They also opened their *samnak chi* for people to stay and helped them to cope with the situation. Most people stayed for about one week.[33] The ordained women in the Asoke movement (the *sikkhamat*s) that I have interviewed did not doubt their ability to assist the victims and were convinced that there was no difference between the *sikkhamat*s' and monks' capacity to help. The ordained women were part of the relief operation that the Asoke organized and were very much appreciated by the survivors.[34]

There are prominent *mae chi*s; for example, Mae Chee Sansanee Sthirasuta, who is a socially engaged and well-known, educated *mae chi* in the Bangkok area. She runs a Buddhist centre that organizes various projects for people in vulnerable situations. She is an exceptional and extremely well-connected *mae chi* with strong bonding, bridging and linking social capital. Both Mae Chee Sansanee and Dhammananda Bhikkhuni, mentioned below, arranged their travel and accommodation privately when they went to Phang Nga.

On 10 January 2005, the Thai female monk Dhammananda Bhikkhuni travelled from Nakhon Pathom, a province near Bangkok, to Phang Nga together with monks, *mae chi*s and laypersons.[35] They went in two vans and during their visit to Phang Nga they stayed at a resort. Dhammananda Bhikkhuni said that she was deeply impressed by how important it was for the survivors that monks and nuns cared about their situation and how vital monks and nuns were for uplifting them spiritually. Dhammananda Bhikkhuni also mentioned that the female survivors appreciated that they could have a more relaxed relationship with the *bhikkhuni*s and *mae chi*s because they were women and they did not have to keep a physical distance, as they had to do with the male monks.[36]

A *mae chi* who was from the affected area and lived in Bangkok at the time of the tsunami went to her home province to help out. She was about 15 years old when she ordained and during her 19 years as a *mae chi*, she had lived at both temples and "nunneries". She had been in Bangkok for ten years and studied with the Open University and she also studied for a master's degree at the Mahamakut Buddhist University. Prior to that, she had studied at a "nunnery" in Ratchaburi province. She also had experience of living at temples in other provinces. She told me about what she did when she heard about the catastrophe and her experiences in Phang Nga:

> On 26 December 2004 I visited Chonburi province together with another *mae chi*. After lunch, I listened to the radio and people were talking about what had happened in the south. I did not pay much attention to the news. I returned to Bangkok by bus and I participated in the chanting at my temple at 4 o'clock as on other days. The next morning, I was at the university as usual. I heard the news on the bus and at the university the teacher and my classmates talked about what had happened. When I came back to the temple that afternoon, I tried to phone my mother in Phang Nga. I saw that I had several missed calls. One was from a teacher in Phang Nga. I also had missed calls from a student that I had been teaching when I was back in Phang Nga a couple of months earlier. That day, I could not get in touch with anybody in Phang Nga. I watched TV and on the 28th my mother called me. I called the head monk at the temple and he asked me to bring certain things that they could not buy in the market. A teacher asked me to bring milk powder. My friends in Bangkok donated money and I bought what they needed. The same evening, I went by bus to Phang Nga. I arrived at 7 o'clock in the morning and the teacher and students were waiting for me and helped to carry all the things. The temple was packed with people and there was a traffic jam on the one big road. I did not know how many people had died, but I knew that they kept many dead bodies at the temple grounds.[37]

After they had carried all the things to the temple, the *mae chi* told the teacher that she wanted to visit the student who had called her when she was in Bangkok. Her mother had brought breakfast to the temple and wanted her daughter to eat, but the *mae chi* said that she did not have time. She went to look for the student along with the teacher. The teacher had a list with names of the students at the school and she checked whom she could find. At around 11 a.m., they arrived in the nearby province of Ranong where the student lived. The *mae chi* said that everything had been destroyed. She had brought some sweets that she gave to the Muslim students who came to meet her. She also met the student who had phoned her. The student told her that her mother had died and her father was injured. The family was preparing for the cremation at their village temple. The *mae chi* met other *mae chi*s and they had lunch together. Before she went back to her village, she also went to visit a hospital. When the *mae chi* arrived at her village in Phang Nga, her family members were preparing for the cremations of her relatives who had died in the tsunami. The *mae chi* said that in the first ten days, most of her time was spent visiting funerals and cremations. Many of the people who had died in the tsunami in that area were her relatives. During that time she also talked with many people about their loss and suffering.

Despite the lack of a *mae chi*s' section, the *mae chi* stayed for many months at the temple and helped out with donations, talked with survivors and helped them in numerous ways. Her parents' house was not far away and she slept there when the temple was too crowded. She was the only *mae chi* at the temple and groups of monks and *mae chi*s from other areas came to help out for a couple of days. People from Bangkok and other places in Thailand who wanted to donate to the victims, but did not know the affected region, contacted the *mae chi* and she helped them with practical matters and she sometimes accompanied them. The *mae chi* also went with donations to the Moken on Surin island. The *mae chi* explained that in the beginning, it was impossible to plan anything because the temple was so crowded with people. They could only live hour-by-hour and day-by-day.

The *mae chi* told me of the criticism of how the temple distributed donations. She said that people who were not affected by the tsunami drove up to the temple and wanted to load their cars with donations for themselves. When the temple refused to give donated goods to people that were not affected by the tsunami, they criticized the monks and the *mae chi* for not distributing the donations. There was also an article in a newspaper stating that the monks did not allow the distribution of donations.[38]

The monks had meetings about how they should manage the work at the temple, but the *mae chi* usually did not attend. She said that she participated in one meeting after the accusatory article was published and they asked her why she had moved donated goods to the school. She explained that she moved the items because the school had a proper room with a roof that protected the donations from being destroyed. Some claimed that she did not distribute the donated goods, but this was not true. She said that she gave to everybody. In the newspaper the abbot was blamed, but at that time he did not know that the *mae chi* had moved some of the goods so that they would be better protected.

The *mae chi* felt that it was difficult to handle the donations. The temple had regulations to follow about how to distribute the donated goods. The *mae chi* reported that people who came and asked for donations had to follow the regulations but she said some tried to get as much as they could even though they were not affected by the disaster at all. She revealed that some people thought that the monks were much stricter than the *mae chi* and therefore they came to her to ask for goods. If they could not get what they wanted, they blamed the temple for not distributing donations.

In a Muslim village that was partly destroyed by the tsunami, some people who were indirectly affected had turned to a newspaper and complained about the donations from the temple. Following the publication, the temple organized a meeting. At the meeting the abbot said that organizations had trusted the temple to manage donations and it was the temple's duty to carry out the task. The temple could not do as it liked with the donations; they had to cooperate with the donor organizations and follow their policies. If the donors were dissatisfied with the temple's management, they would not engage the temple to handle the donations. The temple was under extreme pressure from taking care of the survivors at the temple, handling dead bodies, funerals and cremations and the donations that would sustain the survivors' existence. Later, Thai government officers helped out and managed the materials that were placed at the school. The donations that the temple distributed were kept in tents. The abbot monk told the affected villagers to take what they needed. The abbot provided for the survivors who lived at the temple and enabled the children to go to school. The *mae chi* said that the abbot was very concerned about the survivors' lives.

The *mae chi* tried to calm the conflict between people from the different villages at Koh Phra Thong. She said that it was easier for her than for the abbot because he was from another province and she was from the area. It was easy for her to talk with the people because she knew some of them and they also knew her and her relatives. The conflicts had old roots and she said that the villagers had had problems for a long time. After the tsunami, villagers from Village One stayed for a short time at the temple and were not involved in the dispute. The conflict between villagers from Village Two and Village Four were serious and people from both villages stayed at the temple. Village Four was totally wiped out by the tsunami and the villagers had lost their houses and all their possessions. Village Two was not that severely impacted, though most of the affected people had lost property. The village leader from Village Four raised a claim to the temple that they should have all the donations because of their severe situation and maintained that the donations should be handled by themselves. The *mae chi* said that the village leader came alone and at that point the villagers did not support him. The *mae chi* explained about the conflict that arose:

> There were many people who had lost a number of family members. Not all villagers stayed at the temple. Some of the villagers from Village Four stayed with relatives on the mainland. They also used to come to the mainland and stay together with the relatives before the tsunami. Many of their family members were missing after the tsunami and they were searching for their bodies.[39]
>
> One evening at 9 p.m. Village Four's village leader stated to the temple leaders that Village Two was not affected and he thought they should go back to the island and it was not right for them to stay at the temple. The same night, some villagers from Village Two left the temple to live on the pier. When the abbot and I heard what had happened we went to the pier and asked them to come back to the temple. After that incident I did not pay much attention to the people from Village Four. I knew many people from Village Four but I did not engage myself in their situation. I talked with the volunteers at the temple about the opinion against the villagers from Village Two and the volunteers thought that we should tell them to come back. I

> went and talked with the leader and with the villagers from Village Two and tried to persuade them to let the women, children and old people come back to the temple because it was inconvenient for them to live on the pier. First only women and the children went back and later also the men. The villagers from Village Four continued spending the days on the island looking for their missing relatives. Finally the two villages were separated and Village Four first moved to the school and later they settled on the pier and they received part of the donations. Village Two stayed at the temple.[40]

The *mae chi* said that the conflict went on for a long time and other people who were not initially involved in the conflict also took part in it. She saw the dispute as a power struggle to gain control of the donations rather than being about the wellbeing of the villagers. She said that the village leaders wanted to manage the donations and they did not care about the NGOs' policies. When Village Four had left, the temple managed to distribute all the donations themselves. When the donations were scarce and there was not enough for all the villagers at the temple, the villagers were able to share.

The *mae chi* said that during the first year after the tsunami she stayed about 80 per cent of the time at the temple in Phang Nga and during the rainy season (*vassa)* she was there permanently. After that she went back and forth between Phang Nga and Bangkok and she usually stayed a couple of days. When she went the first time after the tsunami, she thought she was just going there to deliver all the things she had brought for them from Bangkok to the temple. She did not expect to live there for such a long time but she felt that she had to help them in the crisis situation.[41]

Resident mae chis *in the tsunami-hit areas*

Very few temples and *samnak song*s in the tsunami-hit areas had resident *mae chi*s. Most of the *mae chi*s that I met were elderly. One *mae chi* who lived at Wat Bang Muang was 77 years old and told me that she had been ordained twice. She had financed the building of a *kuti* on the temple grounds and there were also *kuti*s that belonged to laywomen who stayed there on special days when they observed the eight precepts. The *mae chi* said that the women who had the *kuti*s close to hers did not speak to her; she said that they thought that the *mae chi*s worked in order to show off.[42] The *mae chi* who helped out at Wat Pasan said that laypeople in Phang Nga did not, in general, have much knowledge about *mae chi*s' restricted life and she said that they had even offered her food in the afternoon.[43]

The *mae chi* at Wat Bang Muang could not live at the temple after the tsunami because it was crowded with dead bodies. She paid the temple for electricity and water, which is also common at other temples. She went on an alms round every morning and it is quite unusual that a *mae chi* at a monks' temple is allowed to do so. She said she used to wake up at about 4 a.m. and listen to *dhamma* talks on the radio and after that she practised walking meditation outside her *kuti*. The *mae chi* kept the temple tidy and she was a part of the temple's team that organized ceremonies and decorated the temple. She was also active in the temple's arrangement of activities for children and novices. Taking care of practical chores, including cooking, have been the main tasks for many *mae chi*s at temples and this has given them little time for studying and meditation.[44]

A *mae chi* at a Buddhist temple in Ranong province was 20 years younger than the *mae chi* at Wat Bang Muang. This *mae chi* also had to pay for water and electricity. She gave massages and laypeople donated what they thought they were worth. At this temple, the *mae chi*s did not go on alms rounds in the morning and the *mae chi* said that she was cooking when the monks went on their alms rounds.[45]

The day before the tsunami struck, monks and *mae chi*s and some relatives had visited the shore not far from the temple. The *mae chi* said that they had been sitting talking on the beach the same time that the tsunami hit the next day. The following day there were about 30 people at the temple, more people came to them crying and describing the giant waves; they said that things were destroyed. The laypeople who had participated in the ceremony just left the temple to see what had happened to their houses. She especially mentioned one woman who had her child in her arms and said that the village was flooded. The woman and child were covered in mud and their clothes were torn to rags. Some people came without any clothes at all. All were weeping and said that many had died. The woman with the baby asked for clothes and the *mae chi* cleaned the baby and covered it with a towel. She said that she decided to give the woman the clothes she had had before she was ordained. She did not know what happened to them after that because they were Muslims and they do not visit the temple. In that area there are more Muslims than Buddhists.[46]

The temple became crowded with survivors and there were many injured people and corpses. Ambulances came to the temple to collect injured people. The *mae chi* said that she felt confused about the chaotic situation. They performed cremations at the temple but some corpses were taken to other temples for cremation. They received more coffins than they used. They gave coffins to temples that needed them and the villagers made tables from the coffins that were left. The *mae chi* said that all the corpses had to be brought to the hospital in Ranong before being cremated at the temple. In Phang Nga, the bodies were not identified at the hospitals and therefore the cremations could be performed without transferring the corpses to the hospital.[47]

The *mae chi* recounted that after the tsunami, the villagers were extremely troubled and they were very busy visiting different organizations and foundations in order to get donations such as rice, drinking water, dry fish and so on. She was concerned about their inner wellbeing because they had to struggle with their basic needs and they were extremely worried about their livelihoods.[48]

A temple in Kuk Kak district in Phang Nga had a section for female practitioners. A Thai woman who was married to a man from overseas had donated money for a special building for *mae chi*s and lay female practitioners. Thanks to that donation, this temple had several *mae chi*s and also a large room for female visitors. The temple had good facilities for laywomen who came to practise for a short time without shaving their heads (*chi phram*). It provided them with food, accommodation and white robes. Both *mae chi*s and *chi phram*s paid for water and electricity to cover the temple's expenses. The *mae chi* section was led by a *mae chi* who was 67 years old. She had been a *mae chi* for 40 years and had spent most of her adult life studying and practising Buddhism. She was from Bangkok and had ten younger sisters and brothers. She had wanted to ordain when she was only 17 years old, but her parents did not give their consent and so she had to wait. She explained the negative attitude towards women's ordination when she was young and said that if daughters ordained, people could think that they had problems – were broken hearted or badly behaved. The *mae chi* said that her desire to ordain came from her faith in practising and she was sad about the prejudices some people had about ordained women.[49]

She said it took her ten years to change her parents' minds and finally they agreed to her ordination. She explained that for ten years before she ordained physically, she was ordained "in her mind". She received ordination in Chonburi province and she selected a place that was far from her family so she would not be involved with them as she had been when she was a layperson. She said that she had always been interested in learning and at the temple in Chonburi she had the chance to study Pali. She finished the sixth level of Pali studies and had already completed the three basic levels of Buddhist studies before she was ordained.[50] She had been a student at one of the monks' universities and she studied Buddhist philosophy and psychology (*abhitham*) at Wat Phra Mahathat in Bangkok. She found the knowledge that she had gained about the mind and consciousness very useful for her meditation practice. She emphasized the importance of combining theory and practice and she said that she had gained a deeper understanding about the *dhamma.* While she was studying, she also worked for the Thai Nuns' Institution, Sathaban Mae Chi Thai. The Nuns' Institution worked with social issues and the *mae chi* said that women have many problems that they prefer to talk about with other women. The *mae chi* was confident about her knowledge and she helped many women with counselling based on Buddhism.[51]

The *mae chi* said that the main objective of her ordination had always been to eliminate defilement (*kilesa*). When she was 39 years old, she began to practise in solitude. She lived in several caves, in the forest and in the mountains. She preferred to practise alone until she got sick, when she was advised by a monk to live and practise at a monastery instead and followed his advice.[52]

The *mae chi* had lived for 20 years in Phang Nga and she said that most *mae chi*s in that province ordained when they were old enough to hold the precepts and perform chanting in order to receive merits as a preparation for death. The villagers have faith in the *mae chi* and the *mae chi* section at the temple is almost like a centre for women. Women turn to the *mae chi* with all practical matters that concern their relationships with the temple and they also come to her for counselling. The *mae chi* said that the villagers respected her and had faith in her and that the temple also benefited from that.

Before the tsunami, the villagers came to the temple for religious merit making (*tham bun*) on *wan phra*, but the *mae chi* said that after the tsunami the survivors increased their religious practice of *tham bun* and many received temporary ordination from seven days up to one month. The villagers experienced suffering and that made them pay more attention to *dhamma*, said the *mae chi.* Suffering taught them the importance of keeping the precepts. She said that Buddhism makes people kind-hearted and she believed that a crisis could be an opportunity for people to live a better life.

The temple provided accommodation for volunteers and rescue teams who came to work after the tsunami. Initially, some monks left the temple because they could not stand the putrid smell of the corpses. The *mae chi* did not leave the temple. Many people took refuge at the temple and were weeping. The *mae chi* said that she advised them to stop crying and have *sati* (mindfulness or self-watchfulness). She said that the survivors did not need the temple for material support because the government took care of that. The *mae chi* helped them with their minds and invited them to stay at the temple to find serenity. She said that the survivors came to the temple; some became ordained, they held the precepts, they showed respect to the Buddha (*wai phra*) and felt stronger when they left.[53]

Notes

1 I will here call both *wat* and *samnak song* temples.
2 The report was published in the newspaper: "Manager" on 1 February 2005.
3 Information from interviews with local monks.
4 Interview, 14 January 2007.
5 Interview, 4 February 2007.
6 Interview, 14 April 2006.
7 Interview, 18 January 2007.
8 Interview, 14 April 2006.
9 Ibid.
10 Ibid.
11 Ibid.
12 Interview, January 2007.
13 Ibid.
14 Interview, 28 February 2007.
15 Interview, April 2006.
16 Ibid.
17 Ibid.
18 Interview, 3 April 2006.
19 Interview, 10 February 2007.
20 Interviews with Moken and Moklen survivors April 2006.
21 Interview February 2007.
22 Conversation with villagers.
23 Information from the temple's "abbott".
24 Interview with a *mae chi* 21 February 2007.
25 Interviews with local monks.
26 Interview, April 2006.
27 Ibid.
28 Interview in Nam Khem in January 2005.
29 Personal communication January 2005.

30 Newspaper: "Manager" 10 February 2005.
31 Interview, 1 August 2007.
32 That was an repeated answer from the *mae chii*s.
33 Interview, 28 June 2007.
34 Interview, October 2005.
35 A female Buddhist monk, *bhikkhuni/bhiksni* (Theravada/Mahayana tradition), has received ordination through the male and female *sangha*s. The Thai *sangha* does not accept *bhikkhuni* ordination.
36 Personal communication.
37 Interview, 21 February 2007.
38 Ibid.
39 Interview, 21 February 2007.
40 Ibid.
41 Interview, 21 February 2007.
42 Interview, 18 January 2007.
43 Interview, 21 January 2007.
44 Interview, 18 January 2007.
45 Interview, 13 September 2006.
46 Ibid.
47 Ibid.
48 Ibid.
49 Interviews, July 2007.
50 The Buddhist studies are called *nak tham* for monks and *thamma seuksa* for laypeople and *mae chi*s.
51 Ibid.
52 Ibid.
53 Interview, October, 2007.

References

Falk, Monica Lindberg (2007) *Making Fields of Merit: Buddhist Female Ascetics and Gendered Orders in Thailand.* Copenhagen and Washington, DC: NIAS Press/University of Washington Press.

Falk, Monica Lindberg (2008) "Gender and Religious Legitimacy in Thailand". In Wil Burghoorn, Kazuki Iwanaga, Cecilia Milwertz and Qi Wang (eds) *Politics in Asia: Women Manoeuvring within Dominant Orders.* Copenhagen: NIAS Press, pp. 95–119.

Heikkilä-Horn, Marja-Leena (1996) *Santi Asoke Buddhism and Thai State Response.* Turku, Finland: Åbo Akademi University Press.

Jackson, Peter A. (2003) [1987]. *Buddhadasa: Theravada Buddhism and Modernist Reform in Thailand*, Chiang Mai: Silkworm Books.

Kawanami, Hiroko (2007) "The *Bhikkhuni* Ordination Debate: Global Aspirations, Local Concerns, With Special Emphasis on the Views of the Monastic Community in Burma". *Buddhist Studies Review* 24(2): 226–244.

Phang Nga: 1st Year Tsunami Memorial (2005) T&D Studio Phang Nga/EPAC Co Ltd. Phang Nga/Bangkok.

Rojanasunan, Porntip (2012) *Dr Porntip Rojanasunan: The Dead Do Talk.* Singapore: Marshall Cavendish Editions.

Weber, Karl E. (2005) *Tsunami 2004: Nam Chai Thai.* Bangkok: Thai National Identity Board, Royal Thai Government.

5 Communication across boundaries

Ceremonies formed an essential part of the recovery process after the tsunami. The practice for Buddhists to communicate across the boundary between the living and dead became the most important ritual among the surviving relatives of the disaster. This chapter deals with loss, deaths and funerals. Survivors narrate how they handled their loss by using necessary ceremonies and how they tried to control "spirits" and "ghosts" after the catastrophe. More unusual ceremonies became important in the search for missing persons and this chapter shows how "new" rituals can emerge in disaster situations. For Buddhists, funerary rites are the most significant of all the lifecycle rituals and Buddhist practices of transferring religious merits to the deceased are here examined in that local setting. At funerals, acts of sharing religious merit are central and relatives of the deceased make merit in order to ensure that the departed family member will have a favourable rebirth. This is of particular importance when a person has died a sudden, violent and unpredictable death. An unexpected death causes much anxiety and worry. An untimely death is considered to be a bad death and the corpse is dangerous in that the spirit, seen as adhering to the body until cremation, has the potential to be a malevolent "ghost". This chapter also deals with informants' experiences of ambiguous loss. When the missing person is not found, it is difficult to deal with and grieving may not be completed because the facts surrounding the loss are unclear.

Rituals

Rituals provide a framework for communication between the living and the dead and for Buddhists the monks are indispensible in their role of bridging the division between the two realms. Informants explained to me that those who had perished in the tsunami were confused about their existence. People had died suddenly; they were unprepared for leaving their families and friends and had not understood that they were dead. Therefore, it was vital to inform the deceased that they were no longer alive and should not intervene in their families' lives anymore. Buddhist monks led the rituals that conveyed that information to the deceased. Laypeople explained to me that in order to increase the deceased's chances of gaining a favourable rebirth, relatives and friends initiated merit-making rituals in which monks were crucial.

The local laypeople that I interviewed after the disaster understood karma in a more collective way than it is explained in the Buddhist doctrine. Karma means action and is taught as something personal that an individual acquires through deliberate acts (see Harvey 2000; Watts 2009). Most Thai people interpret karma in a more inclusive way and family members join together in merit-making rituals and transfer religious merit to the deceased family member in order to increase the chances for a good rebirth. According to the Buddhist doctrine, "the self" and personal identity are neither permanent nor endurable (Collin 1982). After death, the karmic causation and the dependent origination determine both the rebirth's configuration and the realm in which the rebirth will take place. The general patterns of ritual in relation to changes in social status and position that Arnold Van Gennep (1960) has analysed are relevant to the explanation of the transition of rebirth: from being alive to being dead and being reborn. The intermediate stage between living and dying follows the stages in the rites of passage and is especially dangerous.

After the tsunami, collective rituals became an important way for individual survivors to share and cope with suffering. The rituals were also important as a means of binding communities together. They expressed collective ideas about life and death as well as social and moral orders and a shared sense of meaning and understanding. Feelings of belonging and social solidarity were recreated.

Handling untimely deaths

For Buddhists, funerary rites are of ultimate importance. Death resembles a door, used both as an exit and an entrance, and to be prepared for the transition from one existence to another is considered crucial for the next life. In Thailand, people distinguish between death caused by natural causes, such as sickness and old age, and death caused by accidents or violence. The latter is considered to have the potential to cause more deaths and needs to be treated with special care. Heise describes funeral rites in Laos where drowning is considered to be the most dangerous of the "bad" deaths (Heise 2012: 206). The deaths that are believed to be bad in Laos are also considered as bad deaths in Thailand: accidents, the death of children, women who die in childbirth and deaths that occur violently and suddenly without preparation for passing are dangerous for the living and also unwholesome for the dead person. In Thailand their "spirit" (*winjan*) is considered to be lost and has to be convinced to leave.

In Thai Buddhist thinking, the time between death and cremation is hazardous. During this period the deceased is in a state of limbo, not fully departed from this world and not yet transcended into a new existence. Victor Turner has called this process of separation "liminality" (Turner 1966). He focuses on liminality, which had long before been studied as a phase in crises, lifecycle rites and rites of passage (Van Gennep 1960). In Buddhism, death is the starting point of a transformative process towards a new existence (see, for example, Harvey 2000).

The period between the death and the cremation is the time when the spirit can wander amongst the living. It is feared that the spirit of the deceased will haunt the vicinity and "steal" or possess another person's body. Sudden and/or violent death is believed to have the potential to produce a particularly dangerous and uncontrollable spirit, especially if the victim is young and has not acquired life experience and moral discipline (Tambiah 1970). It is thought that those who die an ill-timed death are in need of special rites to make them accept that they have died, leave the living world and, through the rites, gain merits for a rebirth that is as positive as possible.

The tsunami occurred so suddenly that it left people unprepared for death. In the aftermath of the catastrophe, numerous people claimed to have witnessed or experienced ghosts. The spirits' control over the living through their power over the energies in the world requires respectful treatment. The use of new fields, the clearing of land and the building of new homes or enterprises must be carried out with the consent of the environments' spirits. Buddhist monks are not as vulnerable to the spirits as laypersons, therefore it is the monks who conduct ceremonies and who are the vehicles to convey merits to the spirits (Davis 2012: 61).

In Buddhism, the aim of mortuary rituals is to separate the living and the dead, and the funerary practices symbolize and assist the rebirth of the deceased in an appropriately positive place and position. Funerals for ordinary people are held as soon as possible, usually within a week following death. It is preferable to die at home and the corpse is normally kept at home until it is time for cremation. Each day during that period, family, relatives and friends make merits for the deceased by presenting offerings to the monks. On the day of the cremation it is common that sons, grandsons and great-grandsons of the deceased receive ordination as monks and novices.

Cremation can be a simple or very elaborate ritual depending on the deceased's position in society and how much the family can afford to spend. The transfer of merits by the living to the dead does not stop with mortuary rites, but continues after death. The coffin could be covered with a lid or left open according to whether the family members wish to see the face of the departed before the cremation. The coffins that were used after the tsunami were donated to the temples. They were simple and sometimes the coffins were too small, as the corpses had become extremely swollen. Before the cremations at the temples in the post-tsunami situation, the monks took care of the corpses and shrouded them with a thin, white, special fabric. I did not hear of relatives who wanted to see the deceased before the cremation.

The *pamsukula* (Pali – *bangsukul* in Thai) chanting is an important part of the funeral ceremony in Thailand. The origin and changes of the *pamsukula* in Buddhist funeral rites are examined in several articles in the book edited by William and Ladwig (2012). A monk at a *samnak song* explained to me that *pha bangsukul* is a robe without an owner. He told a story from the Buddha's time about celestial beings who wanted to make merit (*tham bun*) and offer monks proper robes because they only had robes made of small pieces of cloth stitched together and then dyed. The celestial beings wove robes of fine quality and placed them so that any monk was free to take and use them. The monk at the *samnak song* said that monks have a specific vow, which means that when they encounter such a robe and there is nobody who owns it, the monk who finds the garment can use it. He continued by telling me that it is not related to dead people but they believe that it is *tham bun* that is dedicated to those who have died. He concluded by saying that ordaining at the funeral meant that the merit earned by the ordination is given to the deceased person, but the person who ordains must have a pure mind. During cremation, a young man can receive novice ordination without having the ceremony in the ordination hall, which is otherwise necessary. The novice should leave the ordained status after the cremation or the day after at the latest if he does not decide to stay in the robes for a longer time. The monk told me that after the tsunami he saw some mourners leaving the novice state too early, immediately after the cremation had begun, and that is not wise.[1]

After death, the dead will be reborn in accordance with the moral quality of past deeds – karma. Ghosts are, from a doctrinal perspective, often the outcome of a lack of *bun* – bad karma – and are associated with bad deaths. Persons who have died a bad death are denied the passage to a new realm and are caught in between the worlds.

Funerals and other religious ceremonies involve acts of giving and transferring merits. The distribution of merit occurs regularly at the end of ritual acts of merit-making, whereby the merit-maker pours water and asks the Goddess of the Earth to carry the merit to all sentient beings (Keyes 2000: 119). At funerals, acts of sharing merit are vital and relatives of the deceased make merit in order to ensure that the departed relative will have a good rebirth. That is especially important when a person has died an unexpected, unpredictable death.

Dealing with ambiguous loss

Ambiguous loss, when the dead body is not found, is difficult to cope with and grieving may not be completed because the facts surrounding the loss are uncertain. Abrams (2001) says that family members may become preoccupied with the lost person and are only satisfied when they receive definitive information about him/her. The usual rituals that help to aid people's mourning are unavailable (ibid.: 283). Central to the ability to make sense out of the loss is the capacity to maintain hope that the missing person is alive, to find some way to change and to construct a meaningful narrative (ibid.: 284). A medical doctor who led the mental health team in Phang Nga province confirmed the difficulties for families with ambiguous loss. She emphasized the importance of finding the remains of the missing people, especially children. She said that those who had found their dead children, cremated them and given birth to new children after the tsunami had recovered, but those who had not found their dead often had difficulty accepting and moving on with their lives.[2]

Some people who had lost several relatives delayed the funeral until they found all the missing bodies and then cremated them at the same time. One informant told me that she had a niece who died in the tsunami and that the niece and her dead grandfather were kept in the ground and were later cremated together with their relatives who had also died in the catastrophe.[3]

While searching for their missing loved ones, family members and relatives performed *tham bun* and dedicated the merit earned to those who had died and to those who were missing. A strategy that some used to find missing people was to place photographs of the missing persons at the funerals of their brothers or sisters who had perished in the tsunami. During the *bangsukul* chanting, the dedication of merits also included the missing persons names and those who were missing were asked to step forward and receive *bun* (merits). The informants explained that if the missing persons were already in a peaceful place, they were promised an even more peaceful place and if they were in a place of suffering, they were urged to leave that place and reap the merits transferred by their relatives.[4]

It was common that relatives received messages from those who had died in their dreams. An informant told me of parents she knew who had dreamt that their children did not have any clothes and were freezing. The next day the parents bought clothes and offered them to the monks at the temple. The monks received the clothes and gave them to people who were in need of clothing. The merits were dedicated to the children, who were informed that their parents had offered what they needed. The informant said the parents told her that, the following night, the children appeared in their dreams and showed them that they had received the clothes.[5]

The number of villagers who visited the temples to make merit increased tremendously after the tsunami. A monk said that there were many people who were missing after the disaster and relatives who could not find their loved ones suffered greatly. He told the relatives to make merit (*tham bun*) and send the merits to the spirits. After they had done that, the relatives felt calmer.[6]

Giving is the primary ethical activity learned by Buddhists. Generosity and giving were also important for Mina, who lost several family members in the tsunami. She did not give alms to monks in the morning because they did not walk the alms round in the area where she lived. She assumed that this was probably due to her home's distance from the temple. Instead, she visited the temple to give food to the monks on *wan phra* and she always dedicated the merits earned to her son and her mother.

Mina dreamt about her son; she said that he communicated with her through her dreams and told her what he needed. The morning after such a dream, she prepared donations and offered them to the monks at the temple. In the coming nights she usually dreamt that her son had received what she had provided to the monks. She said that she was not worried about her son but she was worried about her mother. Mina dreamt every night that she was back in her hometown, cooking food for sale together with her mother. To her, the dream communicated that her mother was not at peace. She thought that her mother wanted her to move back to her hometown and take over the family business run by her sister-in-law. But Mina had no such plans; she said that she could not leave until she had found her son's remains.[7]

Mina lost six family members in the tsunami and she said that in contrast to other families, her family could not find their deceased. Therefore, her relatives decided to organize a funeral without the corpses, which I call a "counterfeit funeral" (*ngan sop look*) at the temple in her home province, about a five-hour drive from where she lived. Her family had decided to have a three-day ceremony, which was conducted five days after the tsunami had occurred.

Hoffman and Oliver-Smith (2002: 11) state that when disasters happen, both old and wholly new ceremonies and rituals arise. A "counterfeit funeral" is a rare ceremony and many Thai informants that I interviewed had never heard about such a funeral before (Falk 2010).[8] Anthropologists and religious experts working in Thailand that I have questioned said that they had never encountered such funerals. It is unusual for the body to be missing after death; however, monks and nuns I have spoken to confirm that they have heard of these funerals, so it is not a wholly new ceremony although seldom conducted.

There seem to be no general regulations about how to conduct the "counterfeit" funeral. In Mina's description of the ceremony, she said that they displayed pictures of the missing persons but they had no coffins. They burned pieces of paper with the dead persons' names written on them. A large number of monks were invited to the funeral and all relatives, neighbours and friends attended the ceremony. The aim was to communicate to the deceased that they had died and encourage them to lead their family and relatives to the dead bodies so they could be cremated.

Mina was opposed to having the "counterfeit funeral" because she still had hopes of finding her son alive. Moreover, she did not want to be away from the temples in the tsunami-hit area to which the dead bodies were brought. However, her relatives were persistent and insisted on Mina's participation. They said that they could not carry out the ceremony without her because the ceremony would then be incomplete. Mina bowed to the social pressure and agreed to participate; her main concern was that she would miss three days of searching.[9]

Mina said that she had difficulties concentrating on the ceremonies at the "counterfeit funeral". On the third day, when the ceremonies were over, her relatives brought her back and she resumed the search for the dead bodies. Soon after the "counterfeit funeral", her mother's corpse, together with four of the other relatives' bodies, were found, but not her son's body.[10]

In 2007 there were still many victims missing after the tsunami and there were many stories about how people had found the dead bodies. Some, like Mina's relatives, found the corpses after they organized "counterfeit funerals". Khunyai is an old woman who had five children – three daughters and two sons. She lost her eldest daughter in the tsunami. That daughter was unmarried and Khunyai lived together with her in a town far away from the sea. Her daughter worked at a beach resort and travelled back and forth every day. She was at work on the day of the tsunami. After the disaster, her brothers and sisters searched for her body and waited for the forensic results. After several months a monk advised Khunyai and her family to arrange a "counterfeit funeral".[11]

The "counterfeit funeral" for Khunyai's daughter was conducted differently to the "counterfeit" funeral described above. This funeral was held at home and not at a temple. Here, they had a coffin inscribed with the missing daughter's name whereas in Mina's case they had no coffins, only pictures of the missing persons. At this funeral they placed the daughter's belongings in the coffin and brought flowers that she appreciated and food that she liked to eat. This contrasts with the other case where arrangements of particular food and flowers favoured by the deceased were not made. They invited monks to conduct the ceremony and they also cremated the coffin – a "counterfeit cremation". This was not done in the former case, where instead they burned papers containing the names of the missing persons. It is important to note that there is no substitute for a body, for example an artificial body, in the "counterfeit" funerals.

A few days after the "counterfeit funeral" and cremation, they received a telephone call from Phuket province confirming that the daughter's body had been found and they were asked to come and collect it. It had been kept in one of the large refrigerated containers and its identity was confirmed by DNA. That was seven months after the tsunami and once again they organized a funeral at home. The ceremony lasted three days and afterwards they brought the body to the temple for cremation.

Funerals that were held after the bodies were found followed the general pattern of conducting funeral ceremonies. Nothing was altered in the ceremonies with reference to the preceding "counterfeit funerals". As we have seen in this and earlier chapters, there were of course differences between how funerals were conducted under normal circumstances and how they were performed in the situation after the tsunami.

Supporting the deceased

When I visited Phang Nga a couple of weeks after the tsunami had struck, it was obvious how important photographs of missing family members were to the survivors. They carried these photographs to the temples and when they searched for those missing in the disaster. Families who had lost all their belongings did not have anything as a remembrance of the missing family member and they asked relatives and friends for photographs. The photographs of the dead and disappeared played an almost sacral role for the mourners. The photographs were treated respectfully and were usually placed close to the Buddhist shrines, becoming one of the most precious possessions there. One woman explained to me the importance of also maintaining the order of events after somebody had died. She said that normally after the deaths of parents the children place the photographs of their mother or father on the elevated shelf so as to pay them respect. However, they will not place photographs of younger, deceased persons on that shelf and parents cannot pay respect to their children. The photographs of deceased children could be displayed on the wall, even though this is not so common. Parents usually keep the bones or ashes of their children at the temple and the reason for this is untimely death.[12]

At ceremonies after the tsunami, the number of photographs of the deceased and missing persons was overwhelming. When the first year after the tsunami had passed, the amount of photographs brought to the temples decreased and instead many people held private ceremonies for the deceased at home or on the beach where the body was found.

Tham bun

Many people that I talked with said that Buddhism has become more important to them after the tsunami. They wanted to learn more about Buddhism and, most importantly, they wanted to make religious merit. As already mentioned, the Buddhist survivors communicated with the deceased through *tham bun*, meritorious offerings, and the monks were essential as the vehicles of the communication. Monks whom I have interviewed stated that *tham bun* was the most important ceremony for relatives and for survivors after the tsunami. Relatives worried that the deceased would have an unfortunate rebirth because of the lack of merit and so wanted to offset that by making *tham bun*. Therefore, the number of *tham bun* ceremonies increased tremendously after the tsunami. The monks liken themselves to post offices, explaining that they sent off the dead person's spirit (*winjan*) at the cremation ceremony and transferred and transformed donations from the living to the dead.

The concept of merit (*bun*) is fundamental to Thai Buddhists' daily practices as well as to special ceremonies. On a daily basis, the majority gave alms to the monks who walked a circuit close to their homes. Many of those who could afford to do so donated more than food to the temples in the names of their deceased relatives. Performing the collective ceremonies for the death was also mandatory for those who had not found their missing relatives and the family members participated in the ceremonies at the temples.

Many families, and especially women, were particular about offering food to the monks in the mornings. It was noticeable that more women than men participated in ceremonies at the temples and they usually went to the temple where they had chosen to keep the bones of their dead relatives. Once a year, families usually invite monks to perform a ceremony at home and transfer merit to the dead family members. Inviting nine monks is the ideal and auspicious number and, preferably, the monks are residents at the temple where the bones are kept. It could be a small or large gathering whereby a male layperson is invited to lead the ceremony.[13] His role is to ask for the Buddhist precepts, lead people in the offering ceremony of *sangha than* and arrange the material for the ceremony. At the beginning of the ceremony one of the monks blesses the water. They have a bowl filled with water, a candle and three types of leaves, which are tied together and placed in the water. During the monks' chanting, the leading monk moves the candle over the water and stirs the droplets of melted wax into it. The *sangha than* offering consists of food and money and, ideally, the laypersons that perform *sangha than* wear white. The laypeople who present donations to the monks hold the offering above their elbow and made a vow (*atthithan*). Other people who participate touch the person with the offering and by that act they are united when they send merit to the dead person. Water is used to transfer merit (*rot nam*). They pour water from one container to another and dedicate merit to the deceased person. They place a lit candle in the bowl containing water and this is an important aspect of the ceremony, as the light is believed to guide the deceased.[14]

The relationship between the deceased and their relatives is characterized by ritual exchanges that are based on a moral universe, which constitutes a compassionate community. The monks are essential conveyors in the transference acts that must also include the sincere merit donors. Informants displayed certain flexibility about how often they observed and conducted ceremonies after the tsunami. All of them said that they performed ceremonies on the third, seventh and one hundredth days after the persons' deaths. A few interviewees carried out more ceremonies on the fifth and eleventh days after death. Many went daily to the temple to make merit for the deceased. Almost all people that I interviewed went to the temple for merit-making on *wan phra.*

The one hundredth-day ceremony emphasized by all informants as extremely important. Some families invited monks and conducted the one hundredth-day ceremony at home, but most people held the ceremony at the temple collectively together with others who had lost their loved ones.

A Nam Khem villager, Somchai, held the one hundredth-day ceremony at home. He lost nine family members in the tsunami: his wife, three sons and five grandchildren. After the tsunami he lived together with his youngest son who had survived the catastrophe. Somchai was born in Nakhorn Si Thammarat and he said that he brought his family to Nam Khem to have a better life. However, the opposite happened and almost all of his family died in the disaster; Somchai said that he is now forever tied to Nam Khem. I met Somchai when he was sitting on the small veranda outside his rebuilt house. His house had been completely demolished by the tsunami and he reconstructed it himself with help from volunteers.

Somchai had been a member of the temple committee at Wat Nam Khem, but he had to resign from that commitment before the tsunami hit. He said that he would like to give alms in the morning but that was not possible because the monks did not walk the alms round in his area. He went regularly to the temple, at least once a week at *wan phra*, and he participated in every Buddhist ceremony that the temple organized. He appreciated the ceremonies and the opportunity to meet and chat with the villagers at the temple. However, he also valued the silence at home. Therefore, he invited monks and organized memorial ceremonies at home for his family members that had perished in the tsunami. He said that *tham bun* was the focus of the ceremony and he explained how the white thread (*sai sin*) was used. The thread was attached to the dead family members' bones, and from there the thread went to the photographs of the deceased loved ones, and finally reached the monks' hands. Somchai also said that when they conducted a special, auspicious *tham bun* ceremony at home, the white thread was first tied to a Buddha statue. On completion of the ceremony, he emphasized the importance of pouring the blessed water used in the ceremony under a tree and he stated that if the water was poured on the ground it was useless.[15]

Somchai said that *tham bun* makes people happy. He and his wife liked to make merit and their daughters followed their example. He said that his daughters cooked food and sometimes he donated the food to the monks at the temple if the daughters were unable to go there themselves. Somchai stated that it is not so important how much a person gives but that it is donated with a pure mind. Every night, before going to bed, he chanted and dedicated the merit to his dead family members.[16]

It was difficult for Somchai to find his missing family members after the tsunami. It was also difficult to identity them and they had to use DNA analysis. When the three remaining family members were found they were cremated at the same time. Somchai had made a vow that if he found all of his missing relatives he would ordain as a Buddhist monk and dedicate the merits to the deceased. During the rainy season in 2005 he received ordination at Wat Nam Khem. That was the third time in his life that he was ordained. When he was young he had been a novice for two years (*phansa*), the second time he ordained as a monk and dedicated the merits to his mother and the third time was after the tsunami. His youngest son survived the disaster and his son received ordination during his mother's cremation ceremony and was ordained for five days. Somchai had wanted to stay in his robes for longer but he left the monastic life after becoming sick. He needed proper treatment at the hospital and had been admitted to hospital twice since he disrobed. He said that if he ordains again it would be for a lifetime.[17]

Somchai was certain that ghosts existed and that we are surrounded by dead people's spirits, although they are invisible to us. He said that a long time ago he had seen ghosts three times. They appeared as humans and they were not frightening, and they wanted living people to make *tham bun* for them. Somchai was not worried about his family members who died in the tsunami. He said that he was convinced that they are fine where they are. They did not suffer, but he said that he suffered because he was alive and therefore had the same problems as usual. His family members had received the merits that had been dedicated to them. Somchai's duty was to make merits for his family and he did not feel at ease if he could not fulfil his responsibilities. He told me that he sometimes prepared his grandchildren's favourite food to donate to the monks. When he performed *rot nam* he always said their names and that they should receive merit from the *tham bun.* Somchai never wrote the names on paper as he felt that they received merit when he said their names. The only time he had written their names on papers was at the first and second memorial ceremonies. On these two occasions he did not bring bones or photographs to the ceremonies.

After the disaster, Amoy lived in a tent at the temporary shelter at Bang Muang. Her house in Nam Khem had been repaired but she was afraid of a new tsunami and did not dare to live there. She had leased out her house and was about to rent a property further away from the water. Amoy was born in Chiang Rai in the North of Thailand. Her father had died when she was nine years old and her mother passed away when she was 20 years old. She had lived in Nam Khem for more than 20 years and had been married but had no children. Before the tsunami struck, she had been a dressmaker.[18]

She described how her mother had always encouraged her to go to the temple and she had kept that routine since she was young. Every evening she paid respect to the Buddha, Dhamma and Sangha and recited Buddhist texts at home, and in the morning she gave alms to the monks. There was a monk who walked the alms round in the area where she lived. She said she offered food and sometimes also money. She always transferred merits after the merit-making in the morning. Sometimes she chanted the long version and sometimes the short one and she always poured the water on the grass under a tree.

The Buddhist monks helped her with instructions about how she could communicate with the deceased. She told me that she dreamt that her dead husband did not have any clothes and she talked with a monk about the dream and what she could do to help. The monk asked her to prepare two threads, one white and one red. She brought the threads to the monk and he made a ceremony at the *stupa* where her husband's bones were kept. She said that after the ceremony she had a dream that her husband wore clothes. Amoy did *tham bun* frequently and dedicated the merits to her dead husband. She wrote his name on papers at ceremonies that included the burning of papers with names of the deceased, and she always said his name in merit-making ceremonies. She thought that she did not properly understand how the different existences worked, but a monk had explained to her that dead people lived in a different world and they listened to the names they had when they were alive. The monk had said that the dead eat the food by smelling it. Amoy had lived in central and north Thailand and she compared the different ways of carrying out *tham bun.* At other places she had lived in they did not burn papers as they do in Phang Nga. In Chiang Rai province, they had a tray with food that the dead person liked and they left joss sticks, flowers and clothes that the dead person liked. A small flag with the name of the deceased was put on the tray and the tray was offered to the monks, who then conducted *rot nam.*[19]

Amoy said that after the tsunami there were many Buddhist monks who had travelled to Phang Nga and offered their help. The monks she had listened to came from Surat Thani and from northeastern Thailand and she said that most of them gave very good *dhamma* talks. The monks organized a short-term ordination retreat for women and Amoy received temporary ordination as a *chi phram* three times after the tsunami.[20]

Another woman, Fa, lived in Kura Buri district in Phang Nga and she had also lost relatives in the tsunami. She was a single parent and had a small shop on the main road. She said that she gave alms to the monks in the morning and after that *rot nam*, and mentioned the names of the persons to whom she dedicated the merits (*bun*) in a short chanting. Fa had been a *chi phram* at a local temple when she was younger but she did not receive ordination after the tsunami since she had to support and take care of her child.[21]

Fa's older brother lost his son in the tsunami and he actively helped out at a temple both before and after the disaster. After the tsunami he became more dedicated to ceremonies and he never missed attending the temple on *wan phra*. She said he thought that if he missed giving alms on *wan phra* his son would go hungry. Her brother had organized ordination for his two sons who survived the tsunami but he actually prepared three sets of robes, bowls and other things needed for the ordination. The third set was for his son who was still missing at that time. He donated the third set to the temple at the same time as his two sons received ordination and the temple offered it to somebody who wanted to ordain. On the one hundredth-day ceremony after the tsunami they had not yet found Fa's nephew. Her brother went to a temple for *tham bun* and he brought a photo of his missing son to the ceremony. Right after the ceremony his son's remains was found. Myanmar workers had heard a crying sound and when they localized the sound they saw a uniform and human bones. Her brother went there to bring back what was left of his son. He recognized his son's necklace, watch and work uniform. The funeral was held immediately after they had found his missing son's remains. According to Fa, it was the *tham bun* ceremony that led her brother to his son.[22]

A *mae chi* from a temple in Phang Nga met many parents who had lost children and property in the tsunami. The temple where she stayed was crowded after the disaster, and many survivors came to *tham bun.* The *mae chi* worked hard at the temple. She stated that the *dhamma* provided the best help and the Buddhist practice was a comfort to the survivors. Many came to ask her for help and she encouraged parents who had lost children to be as tranquil as possible because that would affect their children in a positive way. She said that she suggested the parents perform *tham bun* and dedicate the merits to their deceased children so they could receive *bun.* The *mae chi* explained to me that it is possible to see the dead persons while meditating. She said the children's spirits were still there because they did not realize that they were dead. She concluded that nothing is certain and people who are too attached will suffer more.[23]

Right after the tsunami, a small Myanmar Mon temple was established in an area containing many construction sites with a large number of construction workers from Myanmar. The Mon monks cooperated with the Thai temple and both Thai and Myanmar laypeople participated in ceremonies conducted at both temples. The Thai and the Mon monks held morning and evening ceremonies together. The Mon temple had certain ceremonies that were not held in the Thai temples and even though Thai people were not familiar with them they attended the Mon ceremonies and made merit. They mentioned the names of those to whom they dedicated the merits but in the Mon temple they did not burn papers with the names as was done in the Thai temples. Most of the Myanmar people who lost family members in the tsunami were not fluent in Thai and they needed to talk with the Mon monks in their mother tongue. A Mon monk said that many came to the temple as often as they could to hold eight or ten precepts and dedicated the merits earned to the deceased.[24]

Some schools in the tsunami-stricken areas in Phang Nga organized *tham bun* ceremonies and invited monks to the schools on *wan phra*. Pupils, teachers and villagers participated in the *tham bun* activities at the schools and the villagers appreciated the monks visiting the schools, since they lived far away from the temple.[25]

In all the temples, women were over-represented on *wan phra* and there were always more women than men who gave alms to the monks in the morning. A monk that I spoke to said that he thought that was because women in Thailand did not have an equal opportunity with men to ordain. He said that men think that they are able to receive ordination and thereby create much more merit than women can.[26]

Most monks that I have interviewed about *tham bun* after the tsunami considered the ceremony as a positive act for the survivors even though some monks did not believe in the practice of transference of merits to the deceased. The monks from the Asoke group were extremely critical of the great emphasis on merit-making and the many *tham bun* ceremonies that were organized after the tsunami. The Asoke monks said that *tham bun* could be a way for people to find some relief in the very stressful situation after the catastrophe, but they thought it would be much more useful for people to learn how to conduct a livelihood instead of focusing on *tham bun*.[27]

Ordination

As we have seen from several of the survivors' narratives, ordination as a Buddhist novice, monk, *mae chi* or *chi phram* was an important part of Buddhist practice for the survivors after the tsunami. In Thailand it is common that people go to the temples and make vows to ordain. Mina, who we met in earlier chapters, said that she had made a vow that she would be ordained if she found her dead family members. Making a vow is not part of a ritual and it is done individually, and the reason for Mina's ordination was to send religious merit (*bun*) to her mother and son. In January 2005, after she had found her mother's corpse, she received lay ordination as a *chi phram*. A monk from central Thailand had set up a small Buddhist *samnak song* close to the temporary shelter and had organized a ten-day ordination retreat for women. Approximately 50 women participated; most of them received *chi phram* ordination and five received *mae chi* ordination and shaved off their hair and eyebrows.

During those ten days, Mina said she cried constantly and had difficulty focusing her mind during meditation, but at the same time felt very much at peace. She described how she felt so close to her mother and son that it was as if she could communicate with them. She felt as if she was in a "different world". She told me that during that period many monks had travelled to the affected areas with their teachings and that made her feel like she was in the "*dhamma* world" and not in her ordinary existence. She said:

> After the tsunami I have changed a lot. When somebody asks what I want I just say that I want to have a good family. I want my family to be happy. I do not want much, only enough to live.[28]

In Thailand it is common and desirable that both laymen and laywomen receive time-limited ordination. Women often receive *chi phram* ordination without shaving their heads and men receive ordination without the aspiration to pursue a spiritual or monastic career. Mina has plans to be ordained again when she finds her son and when that happens she will shave her head and eyebrows. She stated:

> My husband and I have tried to find the body. My husband has taken a vow that he will ordain and become a monk when the body is found. We have not found the body yet but my husband decided to ordain and stay as a monk for several months and send *bun* to our son before it would be too late.[29]

Since the tsunami, the ordination of men, women and children has been arranged with the specific purpose of transferring merit to those who have died in the disaster. It is common for temples to organize temporary ordination for schoolboys during the summer vacation. The abbot at Wat Pasan said that, after the tsunami, more boys than ever before had enrolled in the temporary ordination that the temple arranged during the summer holiday.[30]

The ordination ceremonies took two days and the whole village was involved in the activities. In the afternoon on the first day, the boys arrived at the temple together with their families. Villagers who had volunteered to cook food also gathered at the temple. The boys were going to have their last dinner for almost three weeks. As novices they should not eat after noon. Theravada Buddhist ascetics are strict about having only one or two meals a day. They can only eat between 7 a.m. and 12 noon. For the rest of the day they are only allowed to drink.

The first phase of the ceremony was held in the *sala* building. The boys were seated in front of their parents or relatives and bowed to their parents, asking forgiveness for all the wrongdoings that they might have done. The monks arrived and the boys bowed three times and then sat on their knees with hands folded in a respectful position as they received the five Buddhist precepts from the monks, who used their special religious fans in front of their faces. Ordination is a meritorious act for all the participants. The monks held a white thread in their hands while they recited Buddhist texts and the abbot sprinkled blessed water over the boys and their families, and any family member who wanted a white thread had one tied around his or her wrist.

The boys were given a large, green leaf shaped like a cone and the family members each proceeded to cut a strand of hair and place it in the cone. The boys walked on their knees to the abbot and he cut some hair from each boy and placed it in the leaf. After that the boys' heads and eyebrows were shaved. They all wore a white folded fabric that looked like a traditional skirt secured with a belt. The boys were in an intermediate state between the lay and ordained realms. The parents and relatives returned home and the boys spent their first night in the temple. The ceremonies continued the following day with a procession of musicians and villagers who danced on the way to the ordination hall (*bot*). A group of men carried the Thai and Buddhist flags and a large photograph of the Thai king. The boys, with shaved heads and dressed in white, walked after them. Each of the boys had three white lotus flowers, three incense sticks and a candle in their folded hands in front of them. The families and relatives followed at the end of the procession, carrying monks' robes, monks' bowls, thermoses, plastic bowls, soap and other necessary and useful things.

The boys were positioned in front of their parents and relatives and received from them the ochre robes. Then the boys walked on their knees to the monks with the folded robes resting on their outstretched arms. The monks took one piece of the robes and placed it around the boys' necks. The boys went outside the ordination hall and changed into the ochre robes and the abbot gave them the ten Buddhist novice precepts, which the novices repeated after the abbot. The families handed over the monks' bowls to the novices and the last phase of the ceremony took place outside the ordination hall. The laypeople stood on each side of the pathway that led from the ordination hall and the novices walked down the stairs from the ordination hall with their alms bowls in front of them while the laypeople placed food in the bowls.[31]

That group of novices were ordained for 20 days and the religious merit that they earned by being ordained was transferred to those family members and/or relatives who lost their lives in the tsunami. After the disaster, a few temples in the area organized temporary ordination for girls and, as mentioned above, visiting monks organized temporary ordination and retreats for women after the tsunami.

Food for spirits and ghosts

The *Sat Duan Sip* festival, also called the festival of the tenth lunar month, is held on the first and the fifteenth day of the waning moon in the tenth lunar month. People in the south of Thailand believe that this is the time when the spirits of their ancestors, particularly ones that have not yet been reborn, will be released from the underworld. It is a traditional merit-making occasion for the benefit of deceased ancestors. I found that the *Sat Duan Sip* festival was considered to be one of the most important ceremonies by the people in the tsunami-hit areas in Phang Nga. *Sat Duan Sip* was celebrated at all temples and *samnak songs*. People went to the temples and offered food to Buddhist monks and dedicated merit to their loved ones who had passed on to the spiritual realm, as well as to the visiting ghosts (*prets*). Two main merit-making ceremonies were held: one on the first day of the festival and one after 15 days, when the spirits of the dead must return to the underworld.

Significant for the *Sat Duan Sip* festival are some special sweets that people make and they are also available at the markets. They are called *Khanon La*, *Khanom Phong*, *Khanom Tom* and *Khanom Joh Hu*. The sweets represent, for example, clothing for the dead, a vehicle that will take the spirits to heaven, money, and a bed and a pillow. The sweets were offered to the *prets* at the festival.

On the first day of the festival people arrived at the temples at about 8.30 a.m. wearing colourful clothes. They brought two types of food: one for the monks and one for the *pret*s. The special sweets produced for the *pret*s were made of steamed glutinous rice wrapped in banana leaves and another special kind of sweets looked like yellow threads. The threads were made to fit the *pret*s' mouths, which are tiny and only wide as a piece of thread.

The food for the monks was placed in the ceremonial hall and the sweets for the *pret*s were placed outdoors. On the floor outside the ceremony hall there were three black alms bowls on silver plates. White sheets of paper were attached to each bowl with a text that said that the ceremonial rite performed by the monks was done in honour of the dead (*bangsukul*): One for dead people, one for the very tall and hungry *pret* and one for deceased monks. Laypeople who had gathered put money into each bowl. They also wrote deceased people's names on paper and put the papers into the bowl marked for the dead. People went outside the ceremony hall and donated money in the bowls. Some also donated funds by fastening money to the so-called money trees. The temple committee had said that this year they invited people to donate money on this occasion because they were going to spend all the donations on building an ordination hall.

At Wat Bang Muang the festival was organized outdoors. There was a long table for families and relatives' photographs of deceased loved ones. Another large table, made of bamboo, was placed on the temple grounds for people to leave the sweets for the *pret*s. The Moken people gathered around that table. Villagers said that the food was for the *pret* but the Moken would scramble for the food at a certain signal. At some places the alms to the *pret*s were put on a high post so that only the *pret*s could reach them.

At Wat Bang Muang, the monks held a ceremony in one section of the temple grounds and a white thread (*sai sin*) ran from the monks on the raised platform to the bamboo table with the food for the *pret*s. The monks called for the spirits and pulled the white thread, thereby releasing it from the table. That was the sign that the donations to the *pret*s were free for the taking. Poor villagers, children and, in this case, the Moken grabbed the food as quickly as they could.

People who had listened to the monks' *dhamma* talk were transferring merit to their relatives while chanting and pouring water from one container to another. People explained to me that the Moken's competition over food was necessary, as they believed it meant *bun*. If the Moken did not participate in the food competition, their ancestors would be angry with the young people and their revenge would make them ill.

The first day of the *Sat Duan Sip* was also celebrated at Wat Nam Khem and people welcomed the ancestors' spirits (*winjan*). The temple was crowded with people who had gathered and brought food and the special sweets for the *pret*s. They had also brought photographs of deceased family members and relatives that they had placed in the *sala*. A large tray and a monk's bowl were placed in front of the photographs. On the tray, people put sheets of paper with the names of the deceased and the dates they died, and a common date was 26 December 2004. People placed money in the bowl and a paper with the text "*bangsukul*" written on it was attached to the bowl. A monk's bowl with a layperson's name written on it was placed separately from the other items, and money deposited in that bowl was alms to the monk who held the ceremony.

Outside the *sala* were two rows of bowls. In one of the rows people put cooked food into the bowls and in the other row of bowls, people put uncooked rice. People stood in lines to place offerings in the bowls outside the *sala*. A table similar to the one at Wat Bang Muang was placed in the temple's grounds and people brought the special sweets that they had prepared for the *pret*s and they also put flowers on the table.

A white thread was visible running from the middle of the table into the *sala* where the monks held the ceremony. When the monks pulled the thread the rush for the donations on the table was quickly over. Compared with Wat Bang Muang, the people who had gathered around the table outdoors were a mix of people from the village and Moken people. For example, a fisherman from Nam Khem said he wanted a flower from the table to have on his boat for good luck. When the ceremony was over and the table was empty, people left the temple and returned home. Some villagers who had taken food from the table placed it outside the temple. They said that some disabled *pret*s and *pret*s with so much *bap* (bad karma) that they could not enter the temple were outside the gates and were hungry and needed food.

On the last day of *Sat Duan Sip*, the ceremony was organized in the same manner as the first day's ceremony. On the final day, there were more bowls with text about *bangsukul* for specific categories of dead people and there was also one bowl for all the *pret*s. The long table with food for the *pret*s was in place and people had food in simple plastic bags and plastic bowls. They set off firecrackers. The laypeople were transferring merit around the table on the ground and they poured water directly on the ground.

A woman from Baan Nam Khem who always participates in the ceremonies at Wat Bang Muang explained that they write names of their dead relatives, friends and also animals on sheets of paper in order to send merit to them. She said that it is a custom and they have done that for as long as she could remember. The sheets of paper with the names were burnt in the presence of the monks and that was performed as the last part of the ceremony. The burning completed the ceremony and was sometimes carried out late in the afternoon or in the early evening. Few laypeople besides those who assisted the monks participated in the burning ceremony. The laypeople had created merits earlier that day by attending the ceremony in the morning and donating food and sweets to the monks and the *pret*s. The woman stressed that during the burning of the sheets of paper, the monks must perform the *bangsukul* chanting and the dead would thereby receive the merits that their relatives and friends had made for them.[32] The *bangsukul* ceremonies were conducted after the donation ceremonies were completed.

Those who keep bones from dead relatives at home brought the bones to the temple and performed the *bangsukul kraduk* (*bangsukul* for the bones) ceremony. The woman who was our guide during the *Sat Duan Sip* ceremony did not keep bones at home – the bones from her relatives were kept at a *stupa* at the temple. The *bangsukul* ceremony performed for her relatives was usually carried out in the *sala* but she said that it could be conducted anywhere at the temple. A string of white thread was tied from the *stupa* to the place of the ceremony.

Chinese vegetarian festival

The first day of the so-called Chinese vegetarian festival and the last day of the *Sat Duan Sip* festival coincided in 2007. That year, the vegetarian festival was held between 11 October and 19 October. Duang, who we met in Chapter 3, was meticulous about observing the vegetarian festival and so were many in the neighbourhood where I lived in Takua Pa district. Duang explained the importance of preparing for the festival by only taking vegetarian food and observing the Buddhist precepts.[33] Every year, people with a Chinese background in Phuket, Phang Nga and Trang provinces take part in the vegetarian festival. Many of the local Thai people are ethnically Chinese. Duang was adopted and brought up by a Chinese family and followed the Chinese traditions. The vegetarian festival includes features of healing and cleansing and was reported to have begun in Phuket a long time ago when an epidemic hit and affected both locals and a visiting Chinese opera company. The local Chinese community performed certain (Daoist) rites to honour and appease the Nine Emperor Gods and the epidemic ended. The modern festival stems from this and is conducted to cleanse people both spiritually and physically, ward off evil from the coming year and bring good luck (Cohen 2001).

Participants in the festival dressed in white and were particular about refraining from eating meat, drinking alcohol, lying or having sexual relations for nine days. In Phang Nga the festival is held in Chinese shrines that open up specifically for this event. I followed Duang's preparations for the festival and attended rituals where the spirit mediums invited "the gods" to possess them. The mediums prepared themselves by living at the Chinese shrine for a period of time before the festival began and they also stayed there during the festival. When the festival was over they returned to their ordinary civil life. Once possessed, the mediums carried out feats including putting swords through their cheeks and walking on hot coals. They did not feel any pain and showed little or no injury afterwards. The ceremonies included a lot of drumming and firecrackers to ward off evil and as part of the spirit-channelling.

The mediums were both women and men. They were invited to people's homes and walked along the streets with their entourages. A family that lived next door to me invited a group of mediums to their home. Three mediums, one woman and two men, together with a group of helpers, came in two pick-up cars to the house that was located on a small road off the main thoroughfare. My neighbour had placed a round table with a red tablecloth outside the house. On the table were two small vases of flowers, two red candles, three large joss sticks, a teapot and three small cups with tea, a pomelo and a pineapple and a bowl with water and leaves for sprinkling the water. The mediums went into the rooms carrying an open pot of fire that they swung inside the house. One of the male mediums used local liquor to enhance and split the fire. The mediums had whips, which they used when they carried out the cleansing rite inside the house to frighten malevolent spirits. The mediums let off firecrackers to protect the house from evil and the sound was extremely loud. My neighbours explained that the mediums were cleansing the house of bad luck.[34]

The evenings' ceremonies were elaborate and held at the Chinese shrine. Mediums provided healing ceremonies for people who were sick. Some mediums gave massage, some wrote prescriptions for Chinese medicine, some wrote Chinese characters on patients' skin. On the vegetarian festival's last day Duang asked the Nine Emperor Gods to protect her family and their boat, and look after them so they could work in peace. Duang said that after the tsunami the numbers of participants in the vegetarian festival had decreased. She believed that the difficult financial situation after the disaster could be one reason and people were spending their money on *tham bun*.[35] Another possible explanation was a recommendation that people who were in mourning should not participate in the vegetarian festival.

Concluding remarks

The focus of this chapter has been on experiences, narratives, interpretations and actions in terms of spiritual beliefs. In disasters, ceremonies and rituals arise – old, wholly new or with new matter in old form (Hoffman and Oliver-Smith 2002: 11). The "counterfeit funerals", described in this chapter, are one example of a rare ceremony that arose from circumstance and became important for those who experienced ambiguous loss. The Buddhist ordination retreats organized by monks formed an important aspect of the recovery process for many female survivors. For them, rituals and ceremonies became a part of their lives in new ways, offering psychological reassurance and religious efficaciousness as well as communication links between themselves and the deceased. Buddhist monks and nuns have been significant in consoling people, explaining life and death, performing ceremonies and restoring continuity and order. The *Sat Duan Sip* festival and the *Vegetarian* festivals are not Buddhist but they are performed by Buddhists in the south of Thailand. The *Sat Duan Sip* festival is performed at the Buddhist temples and the Buddhist monks participate in the ceremonies.

Notes

1 Interview, 28 February 2007.
2 Personal communication, 20 January 2007.
3 Interview, 22 September 2006.
4 Interviews, March 2007.
5 Interview, January 2007.
6 Interview, 1 February 2007.
7 Interview, January 2007.
8 An article with the "counterfeit funeral" ceremony was published in Religion (Falk 2010).
9 Interview, January 2007.

10 Interview, February 2007.
11 Interview, July 2007.
12 Interview, January 2007.
13 Information from *mae chii*s and lay persons.
14 Ethnography from ceremonies visited in Phang Nga in 2006 and 2007.
15 Interview, March 2007.
16 Ibid.
17 Ibid.
18 Interview, 21 March 2007.
19 Interviews, March 2007.
20 Interviews, March 2007.
21 Interviews, February 2007.
22 Interviews, February 2007.
23 Interview, 14 April 2007.
24 Interview, 15 April 2007.
25 Interviews with teachers.
26 Interview, 10 February 2007.
27 Interview, 1 August 2007.
28 Interview, 4 February 2007.
29 Interview, 14 January 2007.
30 Interview April 2006.
31 Ethnography from April 2006.
32 Interview, October 2007.
33 Interview, 26 October 2007.
34 The visit took place on 15 October 2007.
35 Interview, 26 October 2007.

References

Abrams, Madeleine Seifter (2001) "Resilience in Ambigious Loss". *American Journal of Psychotherapy*, 55(2): 283–291.

Cohen, Erik (2001) *The Chinese Vegetarian Festival in Phuket: Religion, Ethnicity, and Tourism on a Southern Thai Island.* Bangkok: White Lotus Press.

Collin, Steven (1982) *Selfless Persons.* Cambridge: Cambridge University Press.

Davis, Erik W. (2012) "Weaving Life Out Of Death: The Craft of the Rag Robe in Cambodian Ritual Technology". In Paul William and Patrice Ladwig (eds) *Buddhist Funeral Cultures of Southeast Asia and China.* Cambridge, New York, Melbourne, Madrid, Cape Town, Singapore, Sáo Paulo, Delhi and Mexico City: Cambridge University Press.

Falk, Monica Lindberg (2010) "Recovery and Buddhist Practices in the Aftermath of the Tsunami in Southern Thailand". *Religion*, 40: 96–103.

Harvey, Peter (2000) *An Introduction to Buddhist Ethics: Foundation, Values and Issues.* Cambridge: Cambridge University Press.

Heise, Ingmar (2012) "For Buddhas, Families and Ghosts: The Transformation of the Ghost Festival into a Dharma Assembly in Southeast China". In Paul William and Patrice Ladwig (eds) *Buddhist Funeral Cultures of Southeast Asia and China.* Cambridge, New York, Melbourne, Madrid, Cape Town, Singapore, Sáo Paulo, Delhi and Mexico City: Cambridge University Press, pp. 193–216.

Hoffman, Susanne and Anthony Oliver-Smith (2002) *Catastrophe & Culture: The Anthropology of Disaster*. Santa Fe, NM: School of American Research Press.

Holt, John Clifford (2009) *Spirits of the Place: Buddhism and Lao Religious Culture.* Hawai'i: University of Hawai'i Press.

Keyes, Charles F. (2000) "The Cremation of a Senior Monk". In F.E. Reynolds and J.A. Carbine (eds) *The Life of Buddhism*. Berkeley, Los Angeles and London: University of California Press, pp. 124–135.

Tambiah, Stanley, J. (1970) *Buddhism and the Spirit Cults in North-East Thailand.* Cambridge: Cambridge University Press.

Turner, Victor (1966) *The Ritual Process: Structure and Anti-structure.* Chicago: Aldine.

Van Gennep, Arnold (1960) [1909]. *The Rites of Passage*. Chicago: University of Chicago Press.

Watts, Jonathan. S. (ed.) (2009) *Rethinking Karma: The Dharma of Social Justice.* Chiang Mai: Silkworm Books.

William, Paul and Patrice Ladwig (2012) *Buddhist Funeral Cultures of Southeast Asia and China.* New York, Melbourne, Madrid Cape Town, Singapore Sáo Paulo, Delhi and Mexico City: Cambridge University Press.

6 Buddhism and ways to recovery

The capacity for recovery includes many aspects and the social, economic, political and cultural surroundings are all of great importance. Adaptive strategies are based on knowledge, skills and access to resources. This chapter, however, focuses on religion's roles in the recovery process and relates the religious answers given by monks to existential needs as well as to what religion offered to people as a way of understanding the world. Survivors turned to the Buddhist monks for explanations about what had happened and in this chapter Buddhist monks and nuns describe how they helped the survivors and those affected by the tsunami to find solace in grief. The chapter contains monks', *mae chi*s' and survivors' experiences of the recovery process. For the majority of Thai people, Buddhism is the basis of explanations about life and death. After the tsunami, people who did not pay much attention to Buddhism before the disaster also turned to temples for ceremonies, counselling, temporary ordination and to listen to Buddhist teachings. The tsunami became an example of the Buddhist teaching about the conditions of life: that life is unpredictable and inevitably involves suffering of various kinds.

Monks' explanations

In Buddhism all natural occurrences are accounted for in terms of the Laws of Dependent Originations (*Paticcasamuppada*). This means that everything is interconnected and everything is relative and conditional; what happens at one moment is determined by what preceded it. Buddhism views man as a part of nature. If nature is destroyed, man cannot survive and living in harmony with the natural is, therefore, significant. In Buddhism the Four Noble Truths are central: (1) the truth of suffering (*dukkha*); (2) the cause of suffering (*samudaya*); (3) the cessation of suffering (*nirodha*); (4) and the path leading to the cessation of suffering (*magga*). Enlightenment is achieved by following the Noble Eightfold Path, which includes the right actions that lead to the development of the mind. Human existence is characterized by impermanence (*anicca*), suffering (*dukkha*) and insubstantiality (*anatta*) (see Harvey 2000). All are interrelated and humans cannot live independently of the natural environment. Hence, destroying nature causes imbalance.

After the tsunami, religious meanings of the tsunami were discussed. In Western countries leaders of many faiths gave their opinions in the media and some made statements that stirred up a debate (Stern 2007: 3). In the Old Testament, God sent "Noah's flood" to punish sinners and drown a fallen humanity. God was willing to wipe out the majority of mankind as a collective punishment, and after the flood the survivors had a fresh start. That pattern comes back in interpretations of contemporary natural disasters (ibid.: 9). Why does God allow disasters to happen to innocent people? Why does God not intervene in catastrophes and save people instead? These are frequently asked questions, but Buddhism has no divine judge that sees that justice is done. Buddhism is non-theistic and does not have a God that punishes people. Instead there are karma, cause and consequences, created by individual people, which affect every person's life. In Buddhism suffering and dissatisfaction are inevitable parts of life and something that we have created by ourselves; there is no room in Buddhism for blaming others (see Harvey 2000; Watts 2009).

The monks had no knowledge about tsunamis. The phenomenon was unknown to them and the monks said that they first had to get information about it before they could explain to the villagers about the earthquake and the tsunami that followed. The monks that I interviewed after the tsunami told me that they offered simple explanations to the villagers about the conditions of life. They taught the belief that some people have conditions for living long lives and some for short lives and that it depended on the individual's karma – the effect of the person's own intentional actions. The monks in the tsunami area said that they did not teach the belief that the tsunami was any sort of punishment. Generally, the monks did not give any Buddhist explanations of the tsunami; they said that it was a "natural hazard". However, the well-known Vietnamese monk, Thich Nhat Hanh, who lives in France, gave a Buddhist speech on 30 December 2004 in which he spoke about a relationship between negative human actions and the tsunami:

> The human species and the planet Earth are one body. I have the feeling that our planet Earth is suffering, and this tsunami is the cry of the Earth as it writhes in pain: a lament, a cry for help, a warning. We have lived together so long without love and compassion for each other. We destroy each other; we abuse our mother Earth. So the Earth has turned its back on us, has groaned, has suffered. The Earth is the mother of all species. We make each other suffer and we make our mother suffer. These earthquakes are bells of mindfulness. The pain of one part of humankind is the pain of the whole of humankind. We have to see that and wake up.
>
> (Thich Nhat Hanh 2005)

This speech about earthquakes as warning signals for people to change their lifestyles and their ways of treating the environment and each other bore no resemblance to how the monks in the tsunami-stricken areas spoke to the survivors. The latter said they focused on giving people hope and encouraged them to concentrate on the present. In conversations with some monks and nuns in other parts of the country, however, I encountered views like Thich Nhat Hanh's. I interviewed a monk in the affected area who said that most people think that the tsunami is caused by nature but he thought differently. He said that it is human beings who make this world heaven or hell and that has to do with individual people's karma. A natural hazard can happen without killing anybody but if people do not have *bun* (religious merit) and are full of greed, they will suffer. He said that it is karma that forces a natural disaster to become serious and it is karma that both influences nature and shortens people's lifespans. The monk did not explain that to the villagers immediately after the catastrophe. The monk taught them about the Buddhist principles in a similar way that he had done before the catastrophe. He considered that his main role was to guide laypeople so they could reduce their suffering and use their wisdom (*panja*) not only to improve their own material standards but also to improve their minds.[1]

However, one local monk in the tsunami area said that he used to tell the villagers that suffering and dissatisfaction would not last forever and that human beings need to find ways out of suffering. He stated that the best way of solving problems is with wisdom and that some people are wiser than others. Another monk in the tsunami area whom I interviewed said that the basic suffering concerns the body. After the tsunami it was important to help people get medical treatment and also to obtain food, clothes and other basic necessities. Initially, when people were still in shock, he did not deal with suffering that concerned the mind. He did not give any explanations; he simply told them that the tsunami was caused by nature. When it appeared that the survivors were recovering, he tried to explain and teach them about *dhamma*, about cause and effect and about *kamma*.[2]

The most common explanation offered by monks in the field on why the tsunami happened was that it was caused by nature. One monk explained to the survivors that there were three components that operated together in the tsunami. The first component was nature (*thammachat*), the second was karma (*kam*) and the third component was the accident (*ubathihet*). The monk said that when these three components coincide, a catastrophe like the tsunami occurs.[3]

A monk in Takua Pa said that he used the Buddhist concept *anicca* (impermanence) and explained to the survivors that everything is in a process of change. He told the villagers that the tsunami was caused by nature and he noted that nobody questioned or doubted that fact. However, the monk was concerned about how people treated nature. He told me that he believed that the tsunami was caused by *anicca* in combination with an over-consumption of natural resources, like the exploitation of fuel, which he believed affected nature badly. The monk urged the survivors to use natural resources wisely and he suggested that they should work individually and together in groups to find ways to reduce consumption.[4]

The Asoke movement, which formed rescue groups and travelled to the area to help after the tsunami, has elaborated strategies that enable its members to live in as environmentally friendly a way as possible. The Asoke monks taught the villagers about recycling and how to handle household waste. In interviews with Asoke monks who participated in the rescue teams in Phang Nga, they explained that mankind is destroying nature and that if people do not realize the harm they are causing the environment, nature strikes back. They believed that in the future, this will affect humanity more seriously than before. The monks predicted that an even worse disaster would occur if food became poisonous and caused diseases such as cancer, heart failure and high blood pressure. The tsunami brought about a large number of deaths instantly but contaminated or poisonous food would cause slow deaths. The Asoke monks said that they did not bring up that issue with the survivors because they considered that this could make them anxious and the post-tsunami situation was not the right time to raise questions about such possible serious threats. Instead they taught the survivors how to live an environmentally-friendly life, how to cook vegetarian food and how to be more self-sufficient.[5]

The Asoke monks mentioned that there are many stories from the time of Buddha about tragedies that could serve as good examples of catastrophes. One of the monks mentioned an example that happened to the Buddha's relatives. Some of them were killed as a consequence of killing fish. However, the monks said that they refrained from telling those stories, because that would presumably upset the survivors and make them worried. They would perhaps blame the tragedy on themselves, since most of the deceased and the survivors worked as fishermen or lived on incomes derived from the fishing industry. The monks therefore focused on things that were encouraging for the survivors. The Asoke monks said that even though they did not know the people before the catastrophe they tried to support them like they were friends or relatives.[6]

Encouragement

Both local monks and those who had travelled to the area to help out after the disaster said that they tried to help people in the initial stages by encouraging them (*hai gamlang jai*). A monk who took care of many survivors after the tsunami said that he divided the explanations into three simple steps. The first and most important step was to find out and understand what had happened. The second step was to accept what had happened. The third step was to adjust the mind to be ready for the next day, the next week, the next month and the next year. He emphasized the importance of helping others because the survivors would gain strength by doing that. This particular monk summarized by saying that solving the inner problems of the mind is the most important and he encouraged them to help themselves as much as possible.[7]

During several years after the tsunami I regularly interviewed and talked with the local monks in the area. One of those monks had been evacuated together with the survivors from their stricken island to the mainland. Initially after the tsunami people, were busy searching for missing relatives and family members. Their village was demolished and all who survived had moved to temporary shelters on the mainland and one shelter, which was located at a temple. Later a new settlement was built on the mainland for the whole village. The temple had also been wiped out by the tsunami waves but a temple was not included in the plan for the new settlement. The villagers built a simple temporary temple because it was important for them to have the temple close to the village and they wanted to give alms to the monks in the morning. The villagers went to the temple almost every day and listened to *dhamma* talks and participated in ceremonies and activities. The temple organized temporary ordination retreats for both women and men and the villagers said that it was an important part of the recovery process. However, they were reluctant to practise meditation. A monk said that it was difficult for them to concentrate, some were afraid of meditating and they were also busy after the tsunami with many practical matters.[8]

A monk at the temporary temple said that he had the impression that it was more difficult for women to recover from the disaster than it was for men. Women were more concerned and worried about their livelihood. The monk had noted that children had a capacity to be engaged in the present with activities such as football and thereby momentarily forget about the catastrophe. The national donor who set up the new settlement had taken on a larger responsibility than most other aid organizations had done. They arranged seminars and field trips and they also offered workshops to support people's livelihoods. The activities were positive for the survivors and the head of the temple said that the villagers came to participate in the temple's activities, especially when they suffered and needed encouragement.[9]

Another well-educated monk who came from Sukhothai province had also moved to live on a small island before the tsunami struck. He had lived at a temple on the island for two years when the tsunami happened. According to him, 72 people died on the island and all the survivors had been evacuated to the mainland. He moved to a small temple close to the temporary shelter where the villagers from the island had taken refuge. He talked with the villagers every day and they stayed at the shelter for about five to six months. He said that, in the beginning, the most important thing was to encourage the survivors so they would gain some hope. Most of the victims were in shock, and he also went to many of the other temporary shelters and talked with the survivors. He said: "We shared experiences and I wanted them to relax. Those who had lost family members wanted to understand why they had died".[10] He said that he offered explanations according to how well the victims felt. For those who he considered to be in better shape, but suffering from having lost family members, he referred to the Buddha's teaching by telling the story of Kisa Gotami, who lost her child (Lanman 2005: 107). The focus was not on the death of the child, but the experiences of loss and grief felt by the mother:

> Kisa Gotami was a poor woman who faced many hardships during her life. She had lost all her family members including her only son, who died from an illness. She went to the Buddha and asked him to cure the baby, who by that time was already dead. The Buddha responded that he could cure him with proper medicine. He said he had all the necessary ingredients to make

> the medicine except for a mustard seed. He would need a very pure mustard seed – so pure that it had to come from a family that had known no death. Kisa Gotami ran to her village to collect the mustard seed, but every home she visited had known the death of a relative at some point. Finally, she understood the real meaning of the mustard seed: it was a spiritual medicine to teach her the deeper meaning of suffering and compassion.

The monk said he had to be careful about the way he talked to people. He explained about the reality of life and that those who have less *bun* have a shorter lifespan. He and all other monks I have interviewed explained that it was important for the survivors to transfer merit to their dead family members and relatives. As has been mentioned in earlier chapters, after the tsunami merit-making ceremonies grew tremendously.

Cause and effect

Karma, in the Buddhist sense, means simply action but the results of actions are often included in the understanding of karma (Buddhadasa Bhikkhu 1988: 18). In Buddhism, the present life is seen as one of a myriad of existences and is the last in a range of lives that have been lived before. The forms of life are considered to be based on karmic qualities and could be relatively pleasant if they exist in the human realm and likewise in heavenly realms, but the opposite is the case if they exist in the realm of animals, the forms of *preta*s and ghostly beings, or in the several hells that are characterized by constant suffering (Harvey 2000: 12). Not one of the existences in the *samsar*ic realms[11] lasts forever and when the existences' "karmic energy" is used up, there will be a new form and existence via new rebirths. The law of karma governs the forms of existence and the law of karma is seen as a natural law like the law of physics. Pleasant and unpleasant rebirths are the natural results of certain kinds of thoughts and deliberate actions. The different types of karma are understood depending on the values of the deeds and on their performer. The final Theravada Buddhist goal is extinction from *samsara.* Good and bad karma are not enough and therefore thoughts and deeds that do not produce any "karmic energy" are needed because they lead to the ceasing of karma (Buddhadasa Bhikkhu 2005: 5).

The law of karma could be understood as an exact and precise form of justice that is not executed by a God, but by what people have brought upon themselves (Watts 2009). Whatever action a person has done in the past will have consequences in the present or on future existences. That leads to a deterministic understanding of karma, which has important individual and social consequences. For example, all suffering experienced by an individual could be understood as something that has to be endured because of the inevitable law of karma based on deeds done previously (ibid.: 15). The results from the law of karma are not negotiable: bad results could however be conquered by wholesome deeds and vice versa. An influential Buddhist monk and scholar has divided karmic results and the effects into different levels: the individual's mental capacities and quality of mind, the individual's physical appearance and behaviour, the material conditions and events in life and the social conditions the individual lives under (Payutto 1993).

In interviews with people in the tsunami area, they referred to karma and lack of *bun* as reasons why some had died and others had survived. Those who lived in poor circumstances used karma to explain their condition and they also used karma to explain how they hopefully could improve their situation. Many had settled in the region in order to create better circumstances for themselves and before the tsunami struck, this had been a good strategy for most people. People accepted wealth and social differences as the result of good karma and a person who was born rich and male would receive social honour and respect. However, the corruption and fraud that occurred after the tsunami were heavily criticized and were constantly described by the local people as unjust and criminal.

A common notion held by local people was that a female birth is lower than a male birth and is understood as a result of bad karma. These notions are in line with ideas representing women as more worldly and attached than men. The renowned feminist scholar Rita Gross has identified three general attitudes towards women in Buddhist texts. She says that the first teaches that female rebirth is a result of negative *kamma*. The second view imagines a Buddha to be male and, therefore, makes male rebirth necessary for women. The third view is that gender is irrelevant for salvation, in the sense that gender is one of the traits of the ego that needs to be transcended (Gross 1993: 115–117). In Thailand, all three attitudes are articulated and accentuated in various contexts. The third view, which says that gender is irrelevant for reaching enlightenment, is often expressed by the *mae chi*s (Falk 2007: 7–8).

After the tsunami, a frequent question that monks had to answer was why some died while others survived. The monks I have interviewed said that they explained to the villagers that a person dies when their *bun* is finished. The religious merits are dependent on the person's thoughts and deliberate actions and some people wanted to know how young children, who had had a very short life could die because of their lack of *bun*. The monks' answer to that was that their karmic status was nothing that had been created in this life but in their earlier existences. They explained that those who died had come to the end of their lifespans, and they would have died under all circumstances. If the tsunami had not struck, they would have died from something else. Hearing that was a relief for many of those who had lost loved ones. The karmic explanation made them think that it did not matter what they had done; they could not have been rescued since they would have perished anyway. Those who had accepted that their family members, relatives and friends had died, instead of blaming themselves for the deaths, directed their energy to help the deceased to reach *nibbana* or to have a favourable rebirth. Several monks died in the tsunami and two of them had lived at a temple on an island, but they were from other provinces in Thailand. One monk was from the south and the other monk came from the northeast of Thailand. A fellow monk who survived was convinced that the monks who perished were led by their karma to the island to die there. He said that karma chose the right time and karma is the best answer to why something happens unexpectedly.[12]

It is common that people perform good deeds for selfish reasons and cling to their self and to their desires. Few people besides monks and *mae chi*s perform wholesome deeds with the aim of realizing life and reaching *nibbana*. A monk explained that in order to reach *nibbana*, a person had to prepare him/herself by practising meditation and keeping the Buddhist precepts. He stated that when people meet a sudden death, like in the tsunami, they still have defilements (*kilesa*) and therefore they cannot reach *nibbana*. The monk stated that the deceased lives were still filled with too much greed, anger and delusion and they were attached to sensual pleasures. He explained that people suffer because of their defilements and attachment. Those who had practised well could realize the situation when the tsunami struck. But, he added, there were very few villagers who had that advanced level of understanding.[13]

After the tsunami, the villagers were concerned about the deceased's spirits (*winjan*s) and they frequently talked with the monks about how to handle ghosts and *winjan*s. After a ceremony in a local temple, villagers approached a monk with questions about their deceased family members' *winjan*. The *winjan* is present but it cannot be seen. A monk explained to villagers that the *winjan* is similar to the wind, something that we cannot see, touch or communicate with. It is only through meditation that a person can be aware of the spirit. When a person dies, the form disappears and the only thing that remains is *winjan*. He went on to explain about the human existence and the five aggregates (*skandhas*): body and mind, feeling, perception, mental formations and memory and consciousness. He told the villagers that they had done what they could through ceremonies, and karma would determine what new existences would take form.[14]

The monk gave practical advice to the survivors on how to behave if they were affected by a natural hazard like the tsunami. He said that if the person does not die instantly he/she should try to be as calm as possible and think about something positive. If the person's mind is calm, clear and in a meditative state it would be possible to reach *nibbana*.[15] But, that would not be possible if the mind was filled with worries. The monk recommended people to prepare themselves and practise before something disastrous happened because that would be beneficial for them both in life and when they die. The monk told people that they should be prepared to die because death could come at any time.[16]

Villagers said that after death it was best to reach *sawan*, a pleasant higher realm with no suffering, and in order to reach that state they had to purify themselves. Some practised meditation with the goal of reaching *sawan*. But meditation was not enough, they must become free from greed, anger and delusion (*kilesa*) and that implied that they must purify their minds and conduct good deeds. The monk commented that most laypeople had forgotten that it is best to strive for *nibbana* because *nibbana* is forever and unlike *sawan* it does not belong to *samsara*. *Nibbana* is a state of mind that is stable and does not move upwards or downwards. Bad deeds are believed to bring people down to hell and good deeds will provide the person with an existence in the *sawan*. The person will stay in the *sawan* until their *bun* is finished and a new rebirth will occur. That is in contrast to *nibbana*, which does not lead to rebirth.[17]

The mind (explained the monks) has the potential to be refined from a crude form, with a high degree of greed and anger, to a mind with no anger and a minimum of greed. Developing the mind is done through practice and it leads to an understanding that the human body does not belong to us. After death, a crude mind will take the form of a *preta* and the developed mind will have the form of a *thewada.* Being generous and making merit is part of the practice. The merit and "power" (*barami*) are accumulated in the present and previous lives affect this and coming lives. The monk explained that merit-making would improve the present life and would also develop the mind.[18]

Making merits

The process of making-merit is complex. The tradition of laypeople providing monastics and temples with food and other material necessities constitutes a bond between the lay and the ordained worlds and is a core practice in Theravada Buddhism. The Buddha recognized the role of merit-making and signified the monks as "fields of merit" (Egge 2013: 20). High-quality merit is considered to require recipients who are separated from the lay realm and who, through strict living, constitute "field of merit". Giving to monks who belong to a *sangha* is traditionally considered to be the most meritorious form of giving. Giving alms (*dana*) is the benevolent action of giving mainly food, drink and other requisites of life, while generosity (*caga*) is the charitable disposition to do good for others. The main purpose of giving *dana* is, according to the Buddhist doctrine, to get rid of one's attachment to things in this world and to be free from greed (*lobha*) (Harvey 2000: 10, 46). The Buddhist *dana* should not be mistaken for a gift. *Dana* implies no reciprocity and what alms may legitimately be given and accepted are carefully defined and categorized. *Dana* brings the laity and the monastics together. The laypeople theoretically expect nothing from the monk in direct return for their alms. *Dana* should be received without destroying it with reciprocity. *Dana* must be given without obligation or duty, in order to differentiate it from exchange (Falk 2007: 141–143). In practice, however, this ideal appears to be readily displaced by a simple pattern of reciprocity whereby offerings are given to monks in reward for their participation in various merit-making ritual occasions and other services. Monks are, as we have seen, giving to the lay community and Buddhist teaching is conceived as highly valuable *dana* given by monks to laypeople.

In the Theravada Canon a number of instances are described where donations are given in the name of a suffering *preta*, with the aim that the *preta* should attain rebirth as a god because of the karmic fruitfulness of the giving (Harvey 2000: 65.) As we have seen in earlier chapters, donations to monks almost always ends with a ritual sharing of the merits earned with others, such as the deceased in the catastrophe. Sometimes the karmic fruits of good deeds are dedicated to *preta*s and other beings that are seen as having less opportunity to do good deeds themselves. In Thai society a structural inequality has resulted in women and men's unequal possibilities to earn merits through ordination. Therefore, a son's novice ordination is also considered important for his mother. There are changes in the Thai spiritual field and in recent decades *mae chi*s have accentuated their religious vocation by refining their practice, upgrading their educational level and drawing a more distinct boundary between themselves and laypeople, thereby becoming "fields of merit" for laypeople (Falk 2007). Young men are seen to gain maturity through ordination and they also demonstrate filial piety towards their parents. Mothers, in particular, gain much merit through their sons' ordination. Thus, neither the status of monk nor the ambition to attain *nibbana* are final goals for most young men. The aim is rather to accomplish their obligation and re-enter the lay world.

Laypeoples' fear of ghosts

In Thai Buddhist culture, ghosts or spirits (*phi*s) belong to the "unseen" world but have their place among people in the community. Forests, water, land as well as families and the community have their *phi*s. Small spirit houses on stilts are visible in almost every property in Thailand. They are built for the spirit of the land (*Phra Phum*) as protection when a new building is erected. A new structure displaces the spirit of the land and there is a great risk that the *Phra Phum* will create bad luck for the property owner. Therefore, the owner builds a small, beautiful spirit house for the *Phra Phum* to occupy. The *Phra Phum* is looked after by daily offerings in the form of flowers, incense and fruit (Dansilp and Freeman 2002). In earlier research, I have met people who blamed their misfortune on their own carelessness in not attending to the *Phra Phum*. After the tsunami, however, no survivor that I talked with blamed the catastrophe on the spirit of the land.

Thai people in general are afraid of ghosts and believe that it is important not to offend them. Dead persons' *phi*s are not supposed to remain among people. The monks' chanting at the funerals serves to remind the deceased that they are dead. The monks tie a sacred thread around the dead person's home to remind the *phi* that they are no longer alive. A violent death is of special danger for the living. Those who perished in the tsunami were not prepared for death. The beaches and where people died, as well as the temples, cemeteries and places where the corpses were kept, were considered as especially dangerous places haunted by *phi*s and there were many stories that frightened people.

Close to the International Repatriation Centre at Mai Khao was a *thipak song* with five resident Buddhist monks. I spoke with one of the monks, who told me that he was from Songkla province.[19] He had been a monk for 13 years and had been at the *thipak song* for a couple of years. Before he came to Mai Khao he had lived at the famous temple Suan Mokh in Sura Thani province.[20] He said he appreciated the silent life at Suan Mokh but moved to Mai Khao because it was even more peaceful there. He described Mai Khao as a very serene place with a lot of forest, which had changed when they moved the identification centre and all the refrigerated containers there.

The *thipak song* did not perform cremation and the bodies were just stored there. The monks held ceremonies and dedicated the merit to the dead. The monks also helped to carry the bodies. He said that the sickly smell from the corpses stuck to his robes and it did not matter how often he washed them the smell did not disappear.[21]

The monks had one meal per day and they went on alms rounds every day except for *wan phra*, when people came to offer food at the temple. The monk told me that laypeople came to the *samnak song* for advice and counselling. Like most people in the area he had no knowledge about tsunamis. When the waves hit he thought it was a car that had crashed. The sound was so loud when the waves were rolling in, that he said it was like an explosion. He said that about 1,000 people perished in the area where the *samnak song* was located. He explained to people that the tsunami was triggered by natural causes and that we must accept karma. The tsunami happened on *wan phra* when people had come to the temple to offer food to the monks and when they returned, their homes were gone.[22]

The monk at the *samnak song* in Mai Khao gave examples of what they had seen after the tsunami. One man came to a car rental service and rented a car, but people could not see any driver when the car was driven on the road. The explanation that people gave was that it was not a living man but a ghost who had rented the car. The monk said that when people asked him about ghosts, he answered that he had never seen a ghost. He said that people had told him that they had seen the monk with a foreign woman, but when they came closer nobody was there so they thought that it must have been a ghost. A soldier said that he had seen many foreign women standing naked waiting for a taxi. The soldier came to the *samnak* and told the monk about what he had seen and donated robes to the monk with the purpose of forwarding the donation to the women.[23]

The monk in Mai Khao revealed that he had actually also seen ghosts, but he did not tell the laypeople because it would have made them afraid. He said that one evening when he visited the toilet, the place were crowded with foreigners, both men and women. He fetched another monk and when they came back to the toilet the foreigners were gone; the monk concluded that they were ghosts.[24]

The Asoke group, who had come to assist people after the tsunami, also tried to help people overcome their fear of ghosts.[25] The monks encouraged them and tried to convince them that they did not have to be frightened. One Asoke monk told me that some people were so afraid of ghosts that they did not dare to work in the rubber plantations because of the dead bodies and the ghosts that were believed to exist there.[26] He said that the villagers were afraid of going to work in the rubber plantations close to the cemetery. They were terrified of ghosts and there were many rumours about malevolent occurrences that were caused by the ghosts. When the Asoke monks learned about the rumours, they decided to prove to the villagers that the claims were false and thus end their fear of dangerous ghosts. In order to show the villagers that the place was harmless, the Asoke monks decided to sleep in the cemetery and the monk said:

> I stayed at the cemetery until the villagers were not afraid anymore. I walked around where the many bodies were temporarily buried and I smelled the formalin. It was so quiet in the cemetery at night. I also slept four nights on the beach at Nam Khem because the villagers said that dead people were crying on the beach at night. I did not hear any crying and I told the villagers that they did not have to worry about ghosts on the beach.[27]

A monk at Wat Bang Muang described how people had seen ghosts that were from foreign countries standing outside the temple and waving their hands as if they would like to hitch a ride with them, but people did not stop and pick them up. The monks said that he had never seen a ghost and did not believe in ghosts; he added that he had been ordained for 17 years but had never dreamt of ghosts. He said that other monks had seen ghosts and he mentioned that newly ordained monks had seen ghosts the night before they went forth and received full ordination. The young men had become ordained in order to make merits and the monk said that if the ghosts did exist, they were not dangerous. Ghosts need merits and people believe that the ghosts will leave them alone if they make merit and transfer them to the ghosts. The monk said he felt sorry for the villagers who were afraid because of all the deaths after the tsunami. He tried to convince them that the dead people could not harm them and took the monks as an example, since they were living at the temple with more than 1,000 corpses.[28] It was not only the villagers who were afraid of ghosts. The military from Nakhorn Si Thammarat province were also scared and they could not sleep at night. The military transferred dead bodies from Wat Yan Yao to Wat Bang Muang; however, some of them did not dare to enter the big refrigerated containers where the bodies were kept. At one point, a woman who worked in the information technology section had to enter first and then the soldiers could do their job, but they were still frightened and the woman had to promise to stay outside the container so they could see her keeping watch (Rojanasunan 2012: 181). Dr Porntip explained that the haunting noises that came from the corpses were gases that escaped from the decomposing bodies (ibid.: 181). She also mentioned an event when the soldiers said that they had been haunted by ghosts. They were convinced that they saw a group of what they first thought were people walking towards their lodging but then the group suddenly disappeared in to thin air (ibid.: 182).

The people I met when I was visiting the affected sites in January 2005 expressed their appreciation of Dr Porntip's work at Wat Yan Yao. Many of them said that they had seen photographs of her with ghosts standing beside her. The volunteers and others who took care of the dead bodies used masks when they worked, but people told me in an appreciative way that Dr Porntip never wore a mask. They said that she showed respect for the corpses and put her hands together in a *wai* before she began examining the bodies. People appreciated her deferential and sincere way of working; they told me that they admired her and that she had gained much *bun*, by carrying out such demanding work.

At a temple on one of the islands, a monk said to the villagers that it would be possible for them to reach "heaven" if they prepared themselves for dying by doing good deeds. He also explained to the villagers that people had different lifespans that were decided before they were born. He gave an example of a person who had a lifespan of 60 years but who died when he was 40 years old; he therefore had 20 years left and during that 20 years, he could not go anywhere, because his *winjan* was stuck. He also said that some people's *bun* was finished when they died, but not in all cases. His explanations deviated from what other monks said to the laypeople. In an interview the monk said to me that he had met many *winjan*s and they came to him and asked for clothes and for *tham bun.* He said that the *winjan*s work in a telepathic way and he had a reciprocal relationship with them. He communicated with the *winjan*s and if they came at an inconvenient time – for example when he had visitors – asking for food or clothes, he asked them to come another time. The monk told people to make merit and to dedicate the merits to the deceased, and he said that the *winjan*s were pleased when merits were dedicated to them. Many villagers were afraid of ghosts and thought ghosts would bring more deaths. The monk allayed their fear and said that a ghost would not harm a human being; that they came to ask for help and that it was important to respond positively to that request. He explained that the ghosts were not dangerous and they wanted people to make *tham bun* and *rot nam*. The monk considered that the best way to help the ghosts was to give food to the monks and chanting (*suat mon*) would also be helpful.[29] During the interview we were sitting on the floor in the temple's *sala*, and the monk said that under the floor where we sat, were the remains of more than 80,000 cremated bodies. The place had previously been used for keeping corpses and it had been an outdoor cremation place. That was during the period when many people from other parts of Thailand and from other countries had settled in Phang Nga to work in the tin mines. Many of the workers came alone and did not have any relatives to take care of them after they had died. The monk said that no ceremonies were performed for them and the corpses were brought to the island and cremated there.[30]

A monk in one of the affected village temples on the mainland close to the sea explained why giving *tham bun* to the monks is so important. He was ordained in the north of Thailand and he used the story of *anisong buat*, "blessings of ordination", which is often read by a newly ordained monk at the end of the ordination ceremony (Keyes 1986: 80–81). It consists of a myth that tells of the great merit gained by a mother who allows her son to become a novice. The legend describes how the parents did not observe the Buddhist precepts and were at first against their son's wish to ordain. Later, the parents were rescued by their son's good deeds (ibid.) The monk emphasized the importance that the monks follow 227 precepts that make them superior to the laypeople and field of merit for laypeople. The laity observes only five or eight precepts and their *bun* is therefore less. The monks warned laypeople not to neglect the precepts.[31]

Ideal actions

Generosity is one of the ideal actions that are stressed in Buddhism and one that most villagers practised, especially after the tsunami. Other ideal actions and behaviour are truthfulness, loving kindness, compassion, equanimity, wisdom and morality. Being generous creates merits (Harvey 2000: 104). Feeding dogs is good, but giving to laypersons creates merits, especially if the recipient's moral standard is high and they are a good person. Villagers were convinced that giving food to the monks is the best action because there is no doubt that it will render them merits. It is more meritorious to give to the ordained than to ordinary people. A monk said that it is beneficial to give to those who are useful and help the society. Both monks and social workers assist society but donations to the monks will give the donor more merits.[32]

The small village temples usually had a very close relationship with the villagers. At one island village temple, there was only one monk who survived the disaster. A temporary "temple" was set up at the mainland close to a new settlement and was run by this surviving monk. Periodically, novices were ordained and lived at the temple. They helped out with the daily duties and received education from the monk. The monk conducted the daily and other ceremonies, and he taught and organized ordination activities. The monk said that some villagers were depressed because they were fishermen. They knew that it is demeritorious (*bap*) to catch and kill fish and other marine animals. The monk explained that *bap* and *bun* have diverse timelines and it depends when bad or good karma was created as to when the outcomes will be manifested. He thought that most villagers had the wrong understanding and believed that they could "wash" themselves free from *bap* by participating in the temple ceremonies. Most of the villagers on the island had always been fishers. The island had not always had a temple and the monk said that the villagers' vulnerable situation and their struggle to establish a temple were reasons why he wanted to do something for that community. After seven years on the island the monk knew the villagers well and he said that he had empathy for the children and for the community. The villagers were loyal and trusted the monk. The monk's intention had been to support the village and help them to improve and develop their minds.[33]

One metaphor a monk used when talking with the villagers about rebirth was that karma was like an undeveloped film and when the right time came for developing, then rebirth occurred and life became clear and visible. The monk referred to the Buddha, who said that it is individual practice that leads the person forward. People will learn about the condition of life and countless lives through practice. He said that the cycles of lives are like a drama without any ending, driven by greed and desire.[34]

A common explanation that the monks gave to the villagers was that they were born to be fishermen because of their karma and that was something that they could not do anything about, but they could influence their karma in this life. The difficult financial situation and the lack of alternative livelihood opportunities were the main reasons for continuing to work with fishing. Some were traumatized by the tsunami and could not work at sea anymore and they faced a difficult future. The monks tried to encourage the villagers to be hopeful about the future and not worry about things that they could not escape or have any influence upon. They told the villagers that they should try to accept their present situation and make the most of it.[35]

A monk told me that when the villagers felt better, he explained to them that they should not strive for rebirth. After the tsunami many parents hoped to meet their children again and therefore wanted their dead children to be reborn as their offspring. The monk said that wishing for that prolonged the suffering and that meant a death with worrying and lingering sentiment. The monk likened that to honey on the tip of the tongue of a poisonous snake. The parent–child relationship would be repeated again and again and birth, death, meetings and departures would cause suffering. A happy reunion seems as sweet as honey, but this honey is contaminated with poison and that makes people suffer when they lose somebody. The monk suggested that they should try to let go and not strive to meet again. He said that Vipassana meditation might help them to discard relationships and remove fetters and that this would lead to a purer and free mind. He explained that a free mind is a great achievement, which has the potential to lead to the understanding of the *samsar*ic circle of being born, getting old, becoming sick and dying and of how to break that chain. The monk said that he explained to the villagers that in order to purify the mind, they must try not to be too immersed in relationships. This clinging would make them suffer in life after life.[36]

The ceremonies that were conducted after the tsunami helped the survivors in the recovery process. A monk said that many ceremonies are performed because people believe in them and he referred to them as cultural practice. The same monk said that monks' intermediate communicative role performed in the ceremonies they conducted for the dead are actually important for the living. *Bun khoson* are meritorious deeds that living people perform for the dead. The monks said that the survivors became happy because they did something for the people that they love.[37]

Dealing with suffering

The Buddhist monks in the area explained that natural hazards take place because of natural causes and that the world is in a state of constant flux. This, they said, is why Buddhism teaches the belief that human life is *dukkha*, "suffering" and dissatisfaction, because dissatisfaction is an intrinsic and unavoidable part of life and human beings are also subject to such instability. Every being experiences *dukkha* regardless of wealth and status. In the immediate aftermath of the tsunami people suffered, were left shocked and numb, and many have said that they felt that they were unable to speak. They wanted only to be silent. The monks' and nuns' efforts were intended to make people understand and to restore hope for the future.

Dissatisfaction, impermanence and emptiness form a basis for the Buddhist ethics. Overcoming dissatisfaction and freeing oneself and others from suffering are driving forces for Buddhist ethical actions. Generosity is a way out of suffering (Harvey 2000: 122). The Jataka legends are commonly accepted as stories told by the Buddha or his disciples. However, many scholars assume that the Jatakas were originally folk tales that subsequently came to be identified as stories of the Buddha's former lives (Keyes 1987: 181). The last ten Jataka stories, written before the birth of Gotama Buddha, are deemed the most important and among them the last legend of Prince Vessantara is considered to be the most important. The *Vessantara Jataka* is widely known and has attained popularity among Theravada Buddhists. That legend is a natural point of departure for understanding the ideal of generosity in Thai society. This story might, together with two other texts, the *Phra Malai* and the *Trai Phum*, be termed "key" texts of Thai cosmology (Keyes 1987: 181). The virtue of generosity is the main theme in the Vessantara story and Vessantara had found a way that led to the end of the suffering that all beings are subject to. Generosity and non-attachment are also central to that story.

The different forms of disasters and the subsequent consequences that people encounter are part of suffering and are difficult to prevent. Enlightened beings have ended their suffering by entering the *nibban*ic state. However, it is believed that if we live a life in line with the Buddhist teaching on the Eightfold Path, based on morality, meditation and wisdom, individuals have a better chance of escaping from disasters. The miraculous way that some people survived after the tsunami was contributed to the law of karma and the effects of meritorious deeds.[38]

Some survivors believed that natural hazards like earthquakes and tsunamis, that we attribute to nature, are perhaps not wholly caused by nature. Human activities have perhaps played a role in triggering such disasters. It is said that when human beings violate the dynamic stability of planet Earth: the oceans, atmosphere, stratosphere etc., it harms humanity. What we call natural disasters are, therefore, believed by some to be largely man-made, caused by the spiritual degeneration of human society (Watts 2012: 198–199).

A highly respected monk in the tsunami-struck area, who had a reputation of having deep Buddhist knowledge, explained to the survivors about suffering's multiple aspects. He used the tsunami to exemplify the different dimensions of suffering and he provided the survivors with perspectives on how to deal with the consequences. The monk told the survivors that it was possible to evacuate and deal with material loss but it was impossible to do anything about the past and about death since that is beyond our human ability. What he suggested that the laypeople should do was to follow the Buddhist precepts and improve themselves morally in order to create better circumstances for themselves in the present and for the future.[39]

The same monk said that he found that the survivors appreciated honest talk and he considered that the laypeople had handled their situation fairly well. He believed that the survivors' faith in Buddhism had contributed to that. The laypeople had been active at the temple long before the tsunami had hit and came regularly for Buddhist talks, counselling and practices. He said that after the tsunami, the laypeople needed encouragement and he visited them often and advised them on how to relinquish suffering. He tried to talk as simply as possible and not use difficult words. In the post-catastrophe situation the survivors had difficulty concentrating and therefore he gave explanations that were easy to understand. He instructed them to help each other and emphasized that they could support each other in a unique way because they shared the same experiences. The monk said that survivors sometimes came to the temple at night to pay respect to the Buddha statue (*wai phra*) and by doing this, they attained some peace of mind. He listened to them and if he could help them materially, he did so. He mentioned that some donated money but he encouraged them to keep the money for their own use. Some insisted on donating money and then the monk used the money for people in need.[40]

The Asoke monks explained to the survivors that life is characterized by suffering and therefore it is something that we cannot avoid. Finding the cause of suffering and following the Buddhist precept is a Buddhist way to reach equanimity and is essential in order not to create more suffering. A local monk from an affected temple in Takua Pa said that it was a basic feeling when the survivors experienced suffering and they referred that to their own demerit (*bap*). They therefore went to give *tham bun*. The monk said that most laypeople do not understand that if they want to get rid of suffering they must try to get rid of their defilement (*kilesa*).[41]

A monk in Kuk Kak said that many survivors were in great pain and suffering as a result of the tsunami and they came to him for counselling. Most of the survivors had lost loved ones, their houses and their means of livelihood. The monk examined life in general together with the survivors. They talked together about suffering from a Buddhist perspective and a theme that he addressed was impermanence (*anicca*). The monk said that some men chose to ordain as a way to conquer suffering. He considered mindfulness to be the best practice and he said that he instructed the survivors to hold the Buddhist precepts, since that was the basis of everything. He said that meditation was wholesome for some survivors and he gave Vipassana meditation instructions to those who were interested. However, concentration was difficult for many and the majority of the villagers were more interested in getting a talisman than practising meditation and holding precepts.[42]

A monk in Takua Pa district who escaped from the waves said that the survivors needed something to do that could keep their minds away from their traumatic memories and the flashbacks that were troublesome. Most of the survivors had problems with their concentration directly after the tsunami. The monk suggested that the government should have arranged practical work. The survivors were restless and it was very positive for their recovery to carry out manual work. Similar to another monk mentioned above, this monk said that he had found that it was more difficult for women to deal with and overcome suffering than men. He ascribed this to the structural order, where women in Thailand have much more responsibily than men and they worried much more than men about daily life and the lack of livelihood. Men had more physical activities to be engaged in but they also had problems with drinking and gambling, which, the monk said, was discouraging to see.[43]

Another monk in the same district, but not at the same temple, had found that women had more faith in Buddhism than men and women used Buddhism as a refuge. He said that when women lost their husbands, they went to the temple but when men lost their wives they went to have a drink. He concluded that women recovered faster than men because women listen to *dhamma* and they have faith in the monks and in *dhamma.* The monk explained that many men had been ordained as novices when they were young and some had been monks. They had experiences from the temple and had seen that some monks did not follow the precepts and did not live an exemplary monk's life. Therefore, when the novices left the monkhood they did not have the same respect for the monks as the women had, who did not have the same experiences.[44]

Do not look back

Monks and nuns met people in severe situations after the tsunami. The ordained helped them with practical matters and with their inner suffering. It was striking that all Buddhist monks and nuns that I interviewed said they never asked the survivors to tell them about what had happened to them. They never asked about how badly the tsunami had affected them and how many people they had lost in the disaster; there seemed to be an unwritten rule about not talking about the past. One monk said that working in the Buddhist way in a disaster situation simply meant that they encouraged people, gave them hope and tried to make them happy. The monks focused on the present and they stated that it is not fruitful to think about the past, since it is impossible to go back in time and alter events.[45] They said that thinking and talking about the catastrophe and all the things and people they had lost became unbearable and increased the suffering. Loved ones, houses, money and livelihoods were lost and the survivors had to gradually accept that. The monks encouraged the survivors to think about positive things; to take a step towards a new life and adapt to the new environment. The temples organized activities and temporary ordination retreats as a principal part of the recovery process. One goal was to help the survivors appreciate life and give them hope for the future. The monks taught them about life and how to live in a positive way. They emphasized that it does not matter if people are rich or poor; everyone can be a valuable person in society.

A monk who had close contact with the villagers from Koh Phra Thong island said that his advice to the survivors was:

> Do not think about what happened in the past. Release it and try to be in the present, we cannot recall the past and that will make us sorrowful. We cannot ask those who have died to come back. If we think about the past we will not succeed. You must fight for the new life. The life on the island is

> gone and there are new circumstances on the mainland, it is not the same society anymore. You have to adjust yourself to the new environment and forget the past.[46]

Before the Asoke group went to the tsunami area, they had set up certain rules about what they should talk about with the survivors. They did not ask the survivors about their personal stories such as whom they had lost in the tsunami, since that was considered detrimental to the recovery process. The Asoke members talked with them about their loss and how to handle their grief only if the survivors themselves brought up the issue. The monks' way of working was contrary to the way many of the aid workers operated. The aid workers asked the survivors about their loss and what they had been through and that was often counterproductive and a backward step in the survivors' recovery.

Notes

1 Interview, April 2007.
2 *Dhamma* is a word with many meanings; here it refers to the Buddhist view of the nature of things and the teachings of the Buddha. *Kamma* (Pali) karma (Sanskrit) refers to intentional or volitional actions.
3 Interview, July 2007.
4 Interview, July 2007.
5 Interview, August 2007.
6 Ibid.
7 Interview, April 2007.
8 Interview, July 2007.
9 Ibid.
10 Ibid.
11 *Samsara*, (Pali and Sanskrit). The round of rebirths, the conditioned world, as contrasted with the unconditioned, *nirvana/nibbana* (Harvey 2000: 437).
12 Interview, September 2006.
13 Interview, August 2007.
14 Observed in April 2006
15 *Niphan* (Thai) *nibbana* (Pali) nirvana.
16 Interview, February 2007.
17 Ibid.
18 Interview, September 2006.
19 Interview, February 2005.
20 Suan Mokh was established by the highly esteemed monk Buddhadasa, see Jackson (2003).
21 Interview, January 2005.
22 Ibid.
23 Interview, September 2006.
24 Interview, January 2005.
25 The Asoke group is a well-established Buddhist sect that has been excluded from the *sangha* as a disciplinary action (Heikkilä-Horn 1996).
26 Interview, 1 August 2007.
27 Ibid.
28 Interview, September 2006.
29 16 April 2006.
30 Interview, April 2006.

31 Interview, April 2006.
32 Interview, March 2007.
33 Interview February 2007.
34 Ibid.
35 Interview, September 2006.
36 Ibid.
37 Interview, February 2007.
38 That was regularly mentioned as explanations by the informants.
39 Interviews, July 2007.
40 Interview, July 2007.
41 Interview, April 2007
42 Interview, June 2007.
43 Interview, February 2007.
44 Interview, April 2007.
45 Interview, April 2006.
46 Interview, 15 April 2006.

References

Buddhadasa Bhikkhu (1988) *Handbook for Mankind.* Surat Thani, Thailand: Dhammadana Fundation.

Buddhadasa Bhikkhu (2005) "Karma in Buddhism: A Message from Suan Mokh". In Jonathan S. Watts, ed., *Rethinking Karma: The Dharma of Social Justice.* Chiang Mai: Silkworm Books.

Dansilp, Tanistha and Michael Freeman (2002) *Things Thai.* Singapore: Berkely Books.

Egge, James R. [2002] (2013). *Religious Giving and the Invention of Karma in Theravada Buddhism.* New York: Routledge.

Falk, Monica Lindberg (2007) *Making Fields of Merit: Buddhist Female Ascetics and Gendered Orders in Thailand.* Copenhagen/Washington: NIAS Press/University of Washington Press.

Gross Rita M. (1993) *Buddhism after Patriarchy: A Feminist History Analysis, and Reconstruction of Buddhism.* Albany, NY: State University of New York Press.

Harvey, Peter (2000) *An Introduction to Buddhist Ethics: Foundation, Values and Issues.* Cambridge: Cambridge University Press.

Heikkilä-Horn, Marja-Leena (1996) *Santi Asoke Buddhism and Thai State Response.* Turku, Finland: Åbo Akademi University Press.

Keyes, Charles F. (1986) "Ambiguous gender: Male initiation in a Northern Thai Buddhist society". In Caroline Walke Bynum, Steven Harrell and Paula Richman (eds) *Gender and Religion: On the Complexity of Symbols.* Boston, MA: Beacon Press, pp. 66–96.

Keyes, Charles F. (1987) *Thailand: Buddhist Kingdom as Modern Nation-State.* London: Westview Press.

Lanman, Charles Rockwell (2005) *Buddhist Legends.* Delhi: Motilal Banarsidass.

Payutto, Prayudh P.A. (1993) *Good, Evil and Beyond: Karma in the Buddha's Teaching.* Translated by Bruce Evans. Bangkok: Buddhadhamma Foundation.

Rojanasunan, Porntip (2012) *Dr Porntip Rojanasunan: The Dead Do Talk.* Singapore: Marshall Cavendish Editions.

Stern, Gary (2007) *Can God Intervene? How Religion Explains Natural Disasters.* Westport, CT, London: Praeger Publisher.

Thich Nhat Hanh (2005) "Embracing Our Pain. How Can We Explain the Asian Tsunami

Disaster and Heal Our Grief?" *Bangkok Post Outlook*, 30 January 2005 (Dharma talk given in Plum Village, France, 30 December 2004).

Watts, Jonathan. S. (ed.) (2009) *Rethinking Karma: The Dharma of Social Justice*. Chiang Mai: Silkworm Books.

Watts, Jonathan. S. (ed.) (2012) *The Precious Life: Buddhist Tsunami Relief and Anti-Nuclear Activism in Post 3/11 Japan*. Yokohama: the Intenational Buddhist Exchange Center (IBEC).

7 Donations, compensations and religious mission

The devastation triggered the country's biggest-ever relief operation. The day after the tsunami, the Thai government launched emergency relief operations, and the general public promptly played a significant role in supporting the government's response. The Thai public's support of tsunami relief efforts was overwhelming, evident in the national donations of basic necessities, such as food and clothing and the Thai people's swift response to the tsunami was widely acclaimed.

As described in Chapter 4, some temples transformed themselves into centres for forensic teams, temporary shelters and morgues. Initially, the government concentrated their efforts on restoring Phuket province, which was the most important of the six affected provinces financially. Thailand did not accept bilateral or multilateral support from other nations, but the influx of international non-governmental organizations (INGOs), faith organizations and other aid providers became huge. The mismanagement and lack of a national strategy in tackling the catastrophe led to inequities in aid distribution, leaving many survivors without support. While some communities were excluded from government relief efforts, others received excessive amounts of donated materials and a lot of money.

Thai volunteers, Buddhist monks and nuns who lived far away from the affected regions mobilized and immediately travelled to the tsunami-stricken areas to help the local monks with their relief efforts. The former Prime Minister, Thaksin Shinawatra, also reacted swiftly to the disaster and went to the affected regions, appearing in the media inspecting the area, donned in a surgical mask and rubber gloves.

The relief aid can be divided into following groups: the Thai government, international governments and non-governmental organizations, religious institutions and individuals. Immediately after the catastrophe, military personnel and professional search and rescue teams from many countries worked alongside their Thai counterparts. Local and international NGOs played an important role in Thailand's post-tsunami relief operation. The Thai Red Cross Society's role in the relief operation was to coordinate and manage some of the international organizations it was connected to. The announcement by the Thaksin government that Thailand would not appeal for international aid, since they had sufficient domestic resources, did not really alarm the survivors. During the first month after the tsunami, people were simply anxious to receive assistance, regardless of the source of that help. This chapter focuses on several kinds of disaster relief and it presents accounts of how ordained and laypeople experienced their encounters with faith-based organizations and international NGOs.

Thailand's relief and recovery operations

The Thai government was recognized for its effective emergency response. Thailand dealt with the disaster in three steps. In the first stage, the priority was to search for, rescue and evacuate survivors, to deliver essential supplies and to identify human remains. In the next stage, the focus was on preventing infectious diseases among those who had survived and in the final stage, the government concentrated on reconstruction and the longer-term recovery of those affected (UN 2005: 23). Thai volunteers played an important role in providing emergency relief and the Thai public contributed generously with relief supplies and worked round-the-clock in the disaster areas.

Disasters and relief management

In the first phase after the tsunami, the focus was on rescuing people and providing medical aid and basic needs. In contrast to the myth that survivors act irrationally after a disaster, it was observed that survivors behaved quite rationally, and they were the first to respond to their needs and those of their neighbours (Fisher 1998: 20). The initial help in disasters comes from family, neighbours, friends, and those who happen to be around (ibid.: 21–22). Local faith institutions are often among the first providers of immediate assistance, although such help is sometimes criticized, particularly when aid is given solely to co-religionists (Wisner 2010: 129).

The Thai government used the military to construct temporary shelters and permanent houses, and the military was also involved in other relief efforts and in the coordination of identifying the deceased. The Ministry of Public Health sent hundreds of medical doctors to Phuket Province, and the Ministry of the Interior set up a Tsunami Relief Coordination Centre in Phuket. Initially, most of the government's efforts were concentrated on Phuket because of the huge economic importance of the lucrative tourist industry based in the province.

Rescue efforts

Directly after the tsunami catastrophe, a meeting was held with cabinet ministers and concerned government officials at the Southern Tsunami Disaster Victim Logistics Centre. The government decided that:

- "The Thai Army should provide a satellite communication truck to support communication in the Phuket area.

- A force of officers and volunteers should be mobilized to organize search parties in the Khao Lak area, Takua Pa District in Phang Nga Province.
- Phuket and other disaster-prone provinces should be equipped with installations to convey warning calls to residents within three hundred metres from the coastline so as to safeguard against after-shock damage.
- The Department of Disaster Prevention and Mitigation should coordinate with the Po Tek Tueng Foundation and other rescue volunteers and search for survivors and corpses in Khao Lak in the morning of 27 December 2004.
- Transportation should be provided by the Royal Thai Air Force by deploying a C-130 aircraft. A Royal Thai Navy cruiser should be deployed to search for the rescue disaster victims, and helicopters should be used for delivery of food and drinking water, as well as transfer of injured survivors.
- Allocation areas defined for ministries to assume responsibility for direct intervention so as to solve problems expeditiously."

("December 26, 2004 The Tsunami on Phuket" 2005)

Phuket, Phang Nga, Krabi, Trang Satun and Ranong provinces were declared emergency disaster areas, and the Assistance Centre for Victims of the Earthquake and Andaman Coastline Tidal Wave was set up at Phuket Provincial Hall.

The Marine Department of the Ministry of Transport provided 15 boats, and vessels were also obtained from the private sector to transport victims in the maritime territory of Phang Nga province. Aboard each rescue boat were medical doctors and nurses ("December 26, 2004 The Tsunami on Phuket" 2005: 50). The Ministry of Public Health established three centres, one each at Wachira Phuket Hospital, Krabi Hospital and Phang Nga Hospital, to treat the injured and drowned. In addition, a "war room" was set up for coordination and planning. Teams of doctors and volunteers from outside the region joined in these efforts. Through a special register, the tracking of volunteers with specialized capabilities was ensured.

The Department of Disaster Prevention and Mitigation ordered inflatable boats, engines and rescue vehicles as well as equipment from Region 4 Prachuap Khiri Khan province, Region 11 Surat Thani province and Region 12 Songkhla province. The department also requested active involvement of personnel from other government agencies, as well as soldiers, police officers, members of charity foundations and staff from provincial sections. The Thai Army, the Thai Navy, the Thai Air Force, the Thai police, Thai Royal foundations, the Thai Red Cross Society and private foundations were alerted and sent staff and materials for immediate help to the disaster area.

In the afternoon of 26 December 2004, the Ministry of Foreign Affairs set up a 24-hour Coordinating Centre to assist foreign nationals affected by the tsunami. Later that same day, a special flight was arranged to take foreign ministry officials and representatives of the diplomatic missions based in Bangkok to Phuket, to provide immediate assistance to the impacted foreign nationals.

Provincial police helped to coordinate searches for disaster victims, together with other government agencies and private sector organizations. Two days after the disaster, the police set up a Disaster Victim Relief Centre in the Provincial Police Region 8 at the Phuket Provincial Police Office. The police operations were focused on crime prevention, especially theft and the looting of disaster-affected properties. Police also supported forces from the Security Volunteer Civil Section. Other responsibilities included safeguarding people against infections caused by decaying corpses and supporting the identification procedure by recording forensically relevant points ("December 26, 2004 The Tsunami on Phuket 2005": 66–67).

Mental health rehabilitation

For surviving victims, psychological therapy was essential. Different units such as the Drug Addiction Prevention and Rehabilitation as well as the Mental Health units of the Phuket Public Health Office, and the Mental Health Department and relevant sections at Thai hospitals, worked in cooperation with the nursing college of the Thai Army and established the Phuket Mental Health Care Centre for Tsunami Disaster Victims. The mission was to mobilize psychiatrists, psychologist teams, nurses and social welfare officers to diagnose and administer mental therapy as well as offer advice to individuals and groups. The centre reported that half of the disaster victims who came for treatment were severely traumatized and could not sleep ("December 26, 2004 The Tsunami on Phuket 2005": 63, 65). A mental health centre near Khao Lak in Phang Nga province opened in March 2005. The teams based at this centre also had mobile teams and worked with rehabilitation support and screened the tsunami survivors. The Mental Health Department (DOMH) also gave information to schools to support the psychological education that was useful in coping with trauma.

Recovery for children

Help for children orphaned by the tsunami disaster fell under the responsibility of three ministries: The Ministry of Education, responsible for the children's education, the Ministry of Social Development and Human Security, responsible for livelihood issues and the Ministry of Public Health, responsible for their physical and mental health. These three ministries provided support to each provincial office. The provincial-level line agencies were responsible for all orphaned disaster victims, including infants, school-age children and university students (Weber 2005: 62). Almost all children who lost parents were cared for by their extended families. The Thai authorities believed that returning to school was crucial for children's psychological wellbeing. School buildings were repaired and temporary classrooms were put in place. More than 75 per cent of schoolchildren in the tsunami-affected areas were able to return to school two weeks after the disaster (UN 2005: 29). The DOMH mapped out a strategic plan focusing on four different types of affected children: those orphaned by the tsunami, those who witnessed the catastrophe, those whose parents had lost their jobs and homes and those who had poor living standards even before the tsunami struck. These four types of affected children were supported by a two-year mental health relief programme. In Takua Pa district, a mental health care centre was established to conduct surveillance on children's mental health and to evaluate their psychological condition after the tsunami. A team of psychologists was mobilized in the affected areas to provide advice to teachers and relatives on caring for the affected children (UN 2005: 30).

Post-tsunami compensation

Compensation was implemented by the Department of Disaster Prevention and Mitigation as follows:

- compensation for the loss of the head of the family and the principal breadwinner – 40,000 baht per person (about US$1,300)
- compensation for severe injuries – 3,000 baht per person
- immediate compensation for injuries resulting in disability – 10,000 baht per person
- funeral assistance – 20,000 baht per person
- occupational equipment – 10,000 baht per family
- kitchen utensils – 3,500 baht per family
- daily necessity items – 300 baht per family
- clothes – 1,000 baht/person
- school uniforms – 1,000 baht per person
- bedding – 500 baht per person.

With regard to compensation for lost fishing boats, the government provided 66,000 baht for a registered boat and 30,000 baht for a non-registered, small-scale fishing boat.

People who became unemployed as a result of the tsunami received 2,000 baht per person and unemployment benefit of 175 baht per person, per day over a period of 30 days.

Compensation of 25,000 baht per child was provided to children who lost both parents and 15,000 baht was paid to the impacted children.

National rescue operations in Phang Nga

At the provincial administrative level, rescue operations were set up to meet the particular disaster situation in the province. The Phang Nga provincial administration formed five teams with clearly defined responsibility areas, as follows:

1. The rescue-and-relief team was the mobile unit comprising rescue experts who were responsible for searching for survivors in ruins, transferring victims to safer ground and retrieving corpses.

2 The aid-to-victims team was responsible for delivering water, food and consumer goods such as daily necessities, as well as allocating and transferring survivors to temporary shelters.
3 The medical team was responsible for administering first aid to injured people and providing the medical treatment of hospitalized patients.
4 The rehabilitation team was responsible for surveying damaged and destroyed structures, and in charge of planning the speedy rehabilitation of public utilities.
5 The security-and-negotiation team was made up of police officers and volunteers patrolling the disaster area to ensure safety and security, direct road traffic, assist evacuation, remove corpses and provide equipment required for post-mortem identification (Phang Nga: 1st Year Tsunami Memorial 2005: 9, 11–12).

The Fishery Research and Development Centre, established in Phuket province by the Department of Fisheries, set up a support unit to coordinate rescue, relief and rehabilitation in the Andaman Sea coastal region. This centre gave high priority to helping the fishing communities in Phang Nga province. In addition, Phang Nga province created two support centres, one to assist fishermen and another to assist aquaculture entrepreneurs. The officers of units at central and provincial levels were mobilized, and 138 experts manned 12 vessels. They searched the seas around the archipelagos for survivors and retrieved victims' corpses (Phang Nga: 1st Year Tsunami Memorial. 2005: 27). Both staff of government agencies and private sector volunteers helped search for survivors and collected corpses in the coastal area of Phuket Island. Staff and volunteers were in charge of transferring badly injured victims by land, sea and air. Seriously injured victims were evacuated by helicopter and other victims were transported by ambulances and cars. Particularly challenging was the rescue of badly injured victims on remote islands. In as far as they were available, elephants, horses, cattle and buffalo replaced vehicles in such taxing terrain. Private sector enterprises mobilized boats and helicopters to support the government's efforts to make the emergency intervention as effective as possible (Weber 2005: 39).

The Thai Navy retrieved corpses that floated in the sea and Japan's National Defence Force supported in search and rescue operations and provided three equipped vessels. The United States, Asian and European countries also sent rescue teams of more than 400 experts and they brought rescue apparatus, tracker dogs and search equipment (Weber 2005: 39). A centre was set up to register persons reported missing and to update lists of missing persons found alive or dead. Volunteers helped in this task as interpreters and translators. The centre's registers of missing people compiled at the Provincial Halls of Phuket, Krabi, Ranong, Phang Nga, Trang and Satun as well as in each and every district office in the disaster areas, were shared to exchange information and coordinate follow-up actions on missing people with the help of hospitals. Relatives of missing persons who were in Bangkok could contact the Residential Registration Office of the Department of Provincial Administration at the Ministry of the

Interior. The search for missing persons was coupled with a DNA identification service (Weber 2005: 39).

National foundations and relief work

The Buddhist Supreme Patriarch of Thailand donated two million baht to support the survivors. Their majesties and the Thai royal family gave 30 million baht through the Rajaprajanugroh Foundation, under royal patronage, to help disaster victims in the six disaster-hit provinces. The royal foundation provided 2,250 care packages of milled rice, dried food and clothing for the relief of the tsunami victims. Of these care packages, 1,000 were distributed at the Takua Pa District Office, Phang Nga province; 750 sets at the Krabi Provincial hall; and 500 sets were given to survivors in Phuket province. Royal assistance delivered by the Thai Air Force consisted of 1,000 care packages, 54 rolls of white fabric required for wrapping corpses and 275 coffins (Weber 2005: 59). Through the disaster relief sections of the Chaipattana Foundation under royal patronage, funds totalling almost 21 million baht were granted to various foundations also under royal patronage for assistance to disaster victims. The Thai Red Cross Society, under royal patronage, sent out requests by e-mail for the donation of blood, medicine, white fabric, paper tissues, canned food, underwear, sanitary pads, shoes, drinking water, blankets, formalin for the handling of corpses and coffins (ibid.: 59).

The Chaipattana Foundation opened a special account named the "Southern Region Restoration Fund" to administer the numerous contributions from benefactors, including both Thai and foreign organizations, donated to support the tsunami relief projects initiated by their majesties and, upon royal command, supervised by the royal family. The Chaipattana Foundation's disaster relief staff was divided into three groups and each group took care of two provinces. Their first assignment was to assess the damage and to appraise aid provided by other organizations, particularly on remote islands with precarious transportation links, so as to make assessments of requirements and to propose ways and means to provide help (Woothisen 2005).

The overriding concern was to enable the survivors in the worst-hit disaster areas to begin rebuilding their lives. The corresponding support measures covered various essential necessities. Their majesties and the members of the royal family granted scholarships to children who had lost parents or guardians. Emphasis was placed on the resurrection of fishing as the single most important source of sustenance and cash earnings. Due recognition was been given to agriculture as a vital, complementary source of livelihood.

In some cases reported from Kura Buri district, there were not only people whose houses had been destroyed but also people whose houses were more or less intact but who did not want to continue living in the affected area. Their majesties and the royal family were active in Chaipattana Foundation's decision to purchase land and build new settlements on the mainland. These villages gave people, whose houses were destroyed on Koh Phra Thong and Ban Nam Khem, new homes.

Several private sector foundations and organizations, as well as government agencies, collaborated to help schools and students affected by the disaster. The contributions comprised both aid to meet urgent needs and aid to provide support in the long term. The Thai king personally provided funding for the reconstruction of school buildings (Weber 2005: 61).

Immediately after the tsunami struck, the Phang Nga Chapter of the Thai Red Cross Society mobilized teams and made appeals for help from the general public, business people and government officials to cooperate in cooking food for the victims. The distribution of donations was organized and implemented to deliver consumer goods, medical supplies, milled rice, dried food, clothes, kitchenware, electrical appliances, bedding, utensils and beverages. Later, the emphasis shifted to restoring occupations, mental rehabilitation and housing. These tasks were carried out collectively by the Chaipattana Foundation, the Thai Red Cross Society, the Red Cross Society Federation, the International Red Crescent Society, and other domestic as well as foreign organizations. The Phang Nga Red Cross chapter actively supported such projects as the Phru Tiao Red Cross Village and the Ban Nam Khem Infant Nursing Centre (Phang Nga: 1st Year Tsunami Memorial 2005: 5).

Housing and temporary shelters

Local communities supported the national relief effort. Local temples provided shelters and played an important role for individual survivors in many ways. Temples were, as already mentioned, also a base for the forensic teams that worked with identification and for volunteers who came to help with the crisis. A few weeks after the tsunami, the survivors had moved into temporary shelters built by the government or by private organizations. Several shelters were set up in temple compounds or near temples. The survivors were expected to live there for a short time until permanent houses could be rebuilt. But many spent a long time at the shelters. There was a shortage of funding for houses, and people complained about their size. The survivors said that these houses were too small, built by unskilled workers and therefore of poor quality. Building them took a long time and the ownership of them was not always made clear. Survivors were often afraid of moving to their new houses, but there were also survivors who wanted to rebuild their houses at the same place because that was the most convenient location for fishing.

At the end of February 2005, more than two months after the disaster, the number of people who were displaced by the tsunami disaster numbered around 60,000 (Phang Nga: 1st Year Tsunami Memorial 2005: 33). Six months after the disaster, approximately 7,000 people were living in temporary shelters, and the figures for September showed 2,900 persons still living in shelters (UN 2005: 15). The Thai government had received about 3,350 applications for housing assistance and by November 2005, the government had built 2,700 houses (ibid.: 32).

In Kura Buri district a housing project named: "Temporary shelter construction at Samakkhi monastery", was run by Wat Pasan. Phra Kru supervised the constructions of 30 temporary shelter units. The temporary shelters were built for the survivors in Kura Buri district at a cost of 1.2 million baht, funded by the Hong Kong Red Cross Society. Medical and public health care services were offered at temporary shelters and local public health service centres to those affected by the tsunami. The "Red Cross Project for Physical and Mental Recovery" provided medical supplies, non-addictive drugs and medical equipment and tools. A team visited affected people in Takua Pa district who lived in the shelters at the temples Wat Bang Muang, Wat Yan Yao and Wat Lak Kaen, as well as the shelters in Ban Bang Niang, Khao Lak, Ban Pak Wip, Bang Muang, Ban Nam Khem and Ban Nai Rai. The sponsors included the Sirindhorn Red Cross Office of Thung Song district, Thung Song Hospital and various other organizations (Phang Nga, 1st Year Tsunami Memorial, 2005: 72).

Disputes over land

Since the tsunami, there have been disputes over land and families have faced threats and been denied the opportunity to return to and rebuild their houses. Many of those who lived in houses that have been rebuilt in the same place before the tsunami occurred lived with the fear that a tsunami could happen again. Several years after the disaster, these people still listen constantly to the news and are always prepared to run. Some found it too stressful and, therefore, abandoned their houses and moved to safer ground. But, there were also affected people who wanted to move back and rebuild their houses near the sea. For tsunami-impacted communities that had no land ownership papers or titles, the Thai government announced a policy of providing free housing on a resettlement site about 4–5 kilometres inland from their original seafront locations. The government's effort to deliver quick relief and rehabilitation did not allow for extensive community participation, and the well-intended compromise did not deliver the anticipated result as a large number of impacted people, mostly fishing communities, were not prepared to move to the inland sites away from the sea. In some cases, the displaced communities decided to simply leave the relief camps, go back to their old lands and start rebuilding. This led to bitter conflicts and protests because they obstructed proposed civic projects, zoning plans, commercial exploitation and vested interests (UN 2005: 39).

The impact of the tsunami has led to land disputes between local communities, private developers and local governments. This has become a hindrance to the recovery process. Of the 412 villages affected by the tsunami, 83 villages are facing problems related to insecure land tenure. Ethnic groups who live on "prime real estate" along the coast are especially affected (UN 2005: 38). Communities claim that they settled on the land long before title deeds were issued to either private or public entities, some citing the fact that their burial grounds in the community contain ancestors from two to three generations back. When the tsunami hit, many of these communities were not allowed to return to their original settlements and risked being moved further inland, to live in houses away from the sea and the source of their livelihood (ibid.).

Under Thai law, squatters can apply for a legal title to a plot of land after ten years of continuous occupation. Few succeed in acquiring this title and hundreds of people continue to live on what is technically public land in a kind of legal limbo, without papers and without clear rights. When the tsunami wiped away the homes of these communities, it opened up and aggravated the conflicts and disputes regarding ownership of these communities and who had the right to use the land. Evidence of this is provided by the fact that of the 47 villages destroyed by the tsunami, at least 32 were involved in serious land conflicts, about half of them in the province of Phang Nga. In Ban Nam Khem alone, more than 80 court cases over disputed lands have been filed since the disaster (UN 2005: 39).

A special high-level committee led by the deputy prime minister was set up to deal with the more serious land conflicts, which had arisen in tsunami-hit areas. The committee has become an effective tool in finding pragmatic solutions to serious land conflict cases, which allows the people to redevelop their communities on the same land – or on land very close by. In some cases, the communities have been relocated to nearby government lands with long-term individual leases as nominal rent. In other cases, they have been allowed to rebuild on the same public land they occupied previously, as part of an integrated redevelopment plan. In some instances, the communities have relocated from their old sites, which were destroyed, to nearby private land, collectively purchasing "user rights" to public land to which they will be given a community land title from the Land Department. In several cases the land conflict has been solved through "land sharing", where landowners with title deeds have agreed to share the land with the people who have lived there for years, and had considered the land their own. This has required intense negotiation and sensitive mediation, often provided by the Tsunami Land Tenure Committee (UN 2005: 40).

Non-governmental aid

In the early stage after the tsunami, NGOs and volunteers focused on providing immediate relief and resolving instant problems, such as distributing basic goods and daily necessities. Numerous local, national and international public organizations and volunteers provided assistance to the tsunami-affected communities. Each organization implemented a distinct plan that reflected their own philosophy, determining, for example, which category of survivors should receive aid and what the criteria would be for each type of aid. These factors also varied according to the objectives of each organization. Some coordinated the aid by themselves and worked directly with the communities. Other organizations provided support to the local administration for redistribution, while several others collaborated with government agencies. Some permitted the survivors and communities to have a voice so that they could identify their own needs. For some organizations, it was important to prevent duplication and unequal distribution of relief. Most of the survivors I talked with did not trust the way that local authorities distributed relief aid because of nepotism and other forms of corruption.

Despite tremendous public support, the scale of devastation drove Thailand to open its doors to outside assistance. Although the Thai government announced that Thailand would not appeal for international assistance, the country welcomed technical support from its international partners. As a result of this policy, foreign assistance for tsunami recovery in Thailand began at an early stage of relief operations. A high level of international concern was demonstrated by the considerable number of different sources, which provided international contributions to the tsunami recovery in Thailand. One reason for such strong international sympathy was the fact that one-third of the total number of dead and missing was made up of foreign tourists from nearly 40 different countries. This loss prompted foreign governments and individuals from different parts of the world to offer assistance. The sharing of the loss transformed itself into concrete contributions. It is reported that approximately 45 international partners provided technical support in Thailand, and this support and cooperation translated into nearly 200 different projects (UN 2005).

Distribution of aid

Many national and international donors were successful in distributing relief, but there were also problems with disorganization, mismanagement of aid distribution and individual assistance without the necessary long-term strategies being implemented. This made it difficult for the local victims to begin to fend for themselves. The survivors complained about the lack of coordination amongst many aid agencies and the lack of transparency in how the aid was distributed, which created the suspicion that aid agencies favoured one group of villagers over another. Rumours of irregularities became reasons for tensions between survivors in the community. They complained about certain international faith-based NGOs that only gave aid if survivors converted to Christianity. Faith-based agencies that used such methods were disliked by the local residents. Villagers told me that these agencies offered people large amounts of money, houses and boats in exchange for their conversion. They added, however, that they were convinced that in most cases the conversion was not genuine and explained that the survivors had agreed to convert only to get the money and material aid.

Some small communities in Phang Nga were not included in the government relief programmes and the survivors ascribed that to incapacity of the members of local administrative councils to manage the aid situation. Many distressed villagers referred to the lack of fairness in the distribution of the aid by local government staff and members of the local administrative councils assigned by the government to act as the main aid distribution units for the tsunami relief efforts. Community members also questioned the transparency and accountability of such aid distribution.

Information about relief assistance was not widely circulated, so displaced people who did not stay in established temporary shelters were not aware of the assistance they were eligible to receive. Much of the government's information was concentrated on temporary shelters, and those who did not receive such information were unable to apply for assistance on time. Many survivors complained that they were excluded from relief operations due to mistakes made by members of the local administrative councils in their areas. In some cases, survivors who had lost their identity cards or related papers in the devastation were not compensated. Without documents such as work permits, birth certificates, and titles to land and property, victims of the tsunami were unable to obtain certain types of aid, secure employment, health care, or receive inheritance from deceased family members. An informant who lived in the worst-affected village in Thailand told me she was lucky because her house was only partly destroyed by the tsunami. However, almost everything in the house was ruined; the fishing boat that was essential for her and her family's livelihood was smashed into pieces and beyond repair and all the fishing gear was gone. She had also kept a large amount of cash in the house because she had planned to go into town and buy new fishing equipment. She never found the money after the tsunami. We talked about aid and compensation and she said:

> I see many people who used to be poor and after the tsunami they have everything. Those people have not bought all the things that they have – they have received them as donations. They also received money but they have spent the money and now they have begun to sell the things they have received as donations. I think that those who provide relief aid must improve how they distribute donations.[1] The screening is very important. Affected persons were at the hospital and they did not receive any donations despite that they had lost everything. Some persons got a lot of donations and they were not affected at all. That is not fair. Some were given three houses although they had never owned a house before the tsunami.[2]
>
> I received 20,000 baht to repair the house but I requested 50,000. First, I wanted to ask for a new house on safer ground but I changed my mind because I do not want to be disappointed. I could see how the distribution was done after the tsunami. Before the temporary shelters were put up, tents were distributed and I found that one family had got two tents and another family had no tent. They are building a new village with 110 houses. About one thousand people had requested a house there. I wonder where all these people come from? Some boats that were donated were of bad quality and were not built for the sea that we have and could not stand the rough sea. Some old boats were donated to us and they would not last for a long time. There are also boats that are nicely built and the distribution has been fair and there was no conflict in the process when the boats were distributed. We have a fine boat now and are fishing.[3]

The local administrative councils played a key role in the government programme that compensated survivors of the deceased, and they also assisted the government in registering people for aid as well as its distribution. To claim compensation, a member of a family was required to submit a house certificate and related documents demonstrating that the deceased family member was the registered head of the household. The government policy assured that the same amount of money was to provide compensation for both male and female heads of households. However, this turned out to be unjust for some households where the "official" head of the household had not been able to perform the role of breadwinner due to illness or disability, and the responsibility had been passed to a female member of the family.

Some people were left behind in temporary shelters because, according to the government, they were not entitled to a permanent house (provided by the government), since they did not have a certificate of house ownership or were in rented accommodation before the tsunami. The government also did not provide permanent houses to those who refused to relocate from villages where they used to live before the tsunami to designated "safer" areas away from the seashore.

Local sources mentioned unequal distribution of aid by local administrative councils, and several villagers complained that the amount of compensation they received was less than promised. In some villages, victims missed out on government aid because they did not know that they had to apply for it. Most of these villagers were poor and uneducated and they were too traumatized by the situation to take any action in a short time frame. They also had to answer questions about their situation and describe their losses and experiences during the tsunami over and over again, all of which increased their suffering. One fisherwoman, who was in great need of help, told me that it was too painful to answer all the questions repeatedly. Therefore, she fled and locked herself in her room at the temporary shelter as soon as she saw somebody with a writing pad and pen, even though she badly needed the aid. She said that she tried to come to terms with the trauma, but it was made worse by the frequent requirement that she report what had happened, how many people she had lost, etc.[4] While the monks and nuns tried to make people understand and tried to restore hope for the future, their efforts were sometimes counteracted by the lack of coordination in relief work. The survivors were forced to compete for aid and donations, and this was unbearable for many survivors.

People complained and expressed doubts about why and how some survivors received more compensation than others. There were reports of unfair and unrealistic evaluations of property damage when similar damage had been assessed differently, with one person receiving higher compensation compared to others. The lack of trust and mismanagement made individual survivors and organizations turn to Buddhist temples for leadership and what they considered to be "justice" in the distribution of aid.

Rules and regulations

A critical issue in the aftermath of the tsunami was the role played by religious organizations in raising money for relief work. Communities in the tsunami-hit areas in Asia were predominantly Muslim, Buddhist or Hindu, due to which religious organizations of all kinds raised enormous amounts of money. Religious groups also went to the affected areas to distribute aid. Among them, there were many evangelical Christian groups, and the survivors said that the Christians tried to exploit the disaster for their own gains, through proselytising; this was a perception that exacerbated existing religious conflicts or created new ones. As they arrived soon after the tsunami, the dominant presence of Christian groups was conspicuous in a region where other beliefs dominated and inter-religious tensions already existed. This fact posed challenges to interfaith relations and underscored the wide range of attitudes adopted by faith-based agencies with regard to humanitarian work. Some groups adhered to the Red Cross Code of Conduct, which prohibits advocacy of religious or political viewpoints during disaster relief activities. However, many Christian groups put a stronger emphasis on winning new converts, believing that relief can be packaged with religion as long as immediate needs are addressed first.

Norms for behaviour in disaster relief

There are established international norms for the behaviour of NGOs while delivering disaster relief in terms of a code articulated by significant aid organizations. The International Federation of the Red Cross and Red Crescent Societies and the International Committee of the Red Cross jointly prepared the Code of Conduct for the International Red Cross and Red Crescent Movement and NGOs in Disaster Relief. It is sponsored by respected organizations such as: Caritas Internationals, Catholic Relief Services, The International Federation of the Red Cross and the Red Crescent Societies, the International Save the Children Alliance, the Lutheran World Federation, Oxfam, the World Council of Churches, and the International Committee of the Red Cross.[5]

The established international Code of Conduct is a voluntary code and, unfortunately, it was not observed by all organizations after the tsunami in Thailand. The unethical behaviour added to the survivors' suffering in an unacceptable way. I believe that taking advantage of people who have survived a life-threatening disaster is a disgrace and contrary to what humanitarian assistance should stand for.

The Code of Conduct states that:

1. The humanitarian imperative comes first.
2. Aid is given regardless of the race, creed or nationality of the recipients and without adverse distinction of any kind. Aid priorities are calculated on the basis of need alone.
3. Aid will not be used to further a particular political or religious standpoint.
4. We shall endeavour not to act as instruments of government foreign policy.
5. We shall respect culture and customs.
6. We shall attempt to build disaster response on local capacities.
7. Ways shall be found to involve programme beneficiaries in the management of relief aid.
8. Relief aid must strive to reduce future vulnerabilities to disaster as well as meeting basic needs.

9 We hold ourselves accountable to both those we seek to assist and those from whom we accept resources.
10 In our information, publicity and advertising activities, we shall recognize disaster victims as dignified humans, not hopeless objects.[6]

Conditional aid in Phang Nga

I was told that, before the tsunami, there were two Christian churches in Phang Nga, but the villagers did not know what kinds of churches, sects or affiliations they were. By 2007, the number had increased to more than 20. The cross on the signs alongside the roads displayed the large number of Christian organizations that operated in the area. Some Christian groups came to provide aid, whereas some came to convert. Christian missionaries were very active in the hardest-hit areas and my informants told me about their experiences of being approached by Christian missionaries. They had been offered money and material aid if they converted, but the villagers said that they did not want to sell their faith and could not imagine not being Buddhists. They knew people who had converted and who had received a lot in return, but as a result they had lost respect and were looked upon as corrupt. This subject was sensitive and the survivors were reluctant to talk about those who had converted.

Some Christian missionaries were persistent in trying to convert people and visited the same villagers several times in order to coerce them into changing their minds. In one village where the temples were completely destroyed by the tsunami, one of my informants told me that a foreign priest had visited him numerous times. The priest had offered him what he understood as a monthly payment if he converted. He had turned down the offer but had heard from other villagers that the priest used the same method on them (Falk 2013).

In my informant's village, one young man had studied at a Christian school and two women were married to foreign men; these three persons were the only Christians before the tsunami. About two years after the disaster, he estimated that 10 per cent had converted to Christianity. It was believed that the motivation for conversion was the aid and financial support they received. My informants found it strange that those who converted were forbidden to listen to the Buddhist monks and partake in Buddhist ceremonies. The conversion had created divisions within the village and lack of trust between the villagers (Falk 2013: 41–42).

The vast majority of Buddhist monks that I interviewed and communicated with in Phang Nga had a relaxed attitude towards conversion. An Asoke monk said that he met some missionaries who had helped affected people. The monk explained that Buddhism does not prohibit anyone who comes to do good deeds. Conversion is not a problem if laypeople find that it is beneficial. Most monks said that the teaching of all religions is a good thing. The Asoke monk added that the present problem is that the followers do not comply with what the prophets teach.[7]

Concluding remarks

The Thai government has been praised for its rapid response and powerful handling of the disaster situation. Providing quick and effective relief was beyond what any government alone could do. The combined support of skills and resources of many groups, individuals, relief agencies and community networks was necessary (UN 2005: 35).

The role of coordination and proper planning of aid contributions, and the screening of distressed people after a catastrophe, are extremely important in order to avoid adding to their suffering. The lack of coordination among the many NGOs in the worst-hit areas caused the survivors to repeatedly talk about their losses. It is not helpful for the survivors if they have to talk about their experiences and feelings at the same time as they are trying to find out what has happened to their families, children, friends and homes (Koenig 2006: 2). In contrast, Buddhist monks and nuns did not ask the survivors about their loss. Instead they concentrated on their life in the here and now and encouraged the survivors to focus on the near future.

On many occasions, relief approaches made by government agencies and some private donors led to problems such as changing the villagers' attitudes to a "collective" approach to solving problems, and some of the relief projects were organized in ways that supported individuals rather than the whole community, which sometimes resulted in community conflict. People in some communities who used to work together stopped doing so and instead began to fight over donations. Villagers in some areas decided to convert to Christianity, not because of faith, but in an effort to gain access to aid money, housing and other necessities. Fisher concludes that the early post-disaster situation is characterized by altruism. However, in later phases, elements of selfish behaviour, such as looting, appear to be common (Fisher 1998: 21).

The tsunami-hit provinces were treated differently by the Thai government and Phuket province received aid more quickly than, for example, Phang Nga province. Complaints about rehabilitation, housing and livelihoods were raised by local people and civil society groups. Before the tsunami, the issue of land ownership was a complex part of how a society was organized and this has a huge influence on how affected people can cope with a disaster situation.

Coordination issues impacted on the difficult and painstaking task of victim identification. The massive workload and the lack of equipment caused delays in identification work, which obstructed the return of bodies to relatives. There were many lessons learned in the post-tsunami rescue and rehabilitation phases. Vulnerable groups were, for example, not especially protected after the catastrophe. Substantially more attention would be required to adequately address the needs of ethnic groups and unregistered migrant workers who may have been largely overlooked during the tsunami-response operation. Issues of land tenure and titles, the special needs of vulnerable communities, better management of natural resources and the environment and better preparedness for a disaster would be improved.

Affected people did not have much of a say in the construction of houses, the planning of compensation schemes and the formulation of disaster preparation plans. In many cases, communities were not involved in the local administrative organizations' decision making in the recovery efforts. It was obvious that the more marginalized communities had been excluded from local government affairs even prior to the tsunami, both because of their lack of understanding of their rights, and because of the passive approach of some local administrative organizations in reaching out to these communities (UN 2005: 37). The situation of migrant workers in Thailand was also precarious before the tsunami and the disaster compounded their problems.

The lack of information and transparency about compensation schemes created tension within the communities and a lack of participation in planning by the survivors caused internal conflicts among community members. If the Red Cross Code of Conduct or a similar code of conduct had been followed, survivors would have been less vulnerable, and they would not have been exploited by organizations that only wanted to provide aid on the condition of religious conversion. Those organizations that were most successful in their efforts had been working with the affected communities before the tsunami or had experience and knowledge about the local setting. They knew the environment and had established relationships with the communities, and they were able to restore hope that the survivors had the capacity to be self-reliant.

Notes

1 Interview, January 2007.
2 Ibid.
3 Interview, 22 September 2006.
4 Interview, October 2006.
5 See www.gdrc.org/ngo/codesofconduct/ifrc-codeconduct.html (accessed 4 June 2014).
6 Ibid.
7 Interview, 1 August 2007.

References

December 26, 2004 The Tsunami on Phuket (2005). Phuket: Phuket Province 2005.

Falk, Monica Lindberg (2013) "Thai Buddhists' Encounters with International Relief Work in Post-tsunami Thailand". In Hiroko Kawanami and Geoffrey Samuel (eds) *Buddhism, International Relief Work, and Civil Society.* New York: Palgrave Macmillan.

Fischer, Henry W. (1998) *Response to Disaster: Fact Versus Fiction and its Perpetuation – the Sociology of Disaster*, 2nd edition. Lanham, MD: University Press of America.

Koenig, Harold, G. (2006) *In the Wake of Disaster: Religious Responses to Terrorism and Catastrophe.* Philadelphia and London: Templeton Foundation Press.

Phang Nga: 1st Year Tsunami Memorial (2005) T&D Studio Phang Nga/EPAC Co Ltd. Phang Nga/Bangkok.

United Nations (2005) *Tsunami Thailand One Year Later – National Response and the Contribution of International Partners*. Bangkok: UN.

Weber, Karl E. (2005) *Tsunami 2004: Nam Chai Thai.* Bangkok: Thai National Identity Board, Royal Thai Government.

Wisner, Ben (2010) "Untapped Potential of the World's Religious Communities for Disaster Reduction in an Age of Accelerated Climate Change: An Epilogue and Prologue". *Religion* (special issue: Religions, Natural Hazards, and Disasters) 40 (2): 128–131.

Woothisen, Onanan (2005) "The Royal Benevolence Wiped Away Tears". Tsunami Chai Pattana Andaman: 22–35. Available at: www.chaipat.or.th (accessed 4 June 2014).

8 Remembrance, commemoration and memorial ceremonies

In post-disaster situations, memorials are indispensable. Memorial ceremonies serve varying purposes, from national recognition of a tragedy to healing an individual's inner wounds. After the tsunami, local memorial ceremonies were important for the survivors. The tsunami became a national catastrophe and people all over the country suffered together with those who were directly affected by the waves. People of all faiths and from all walks of life united in mourning the loss of loved ones and paying them tribute. Remembrance ceremonies are part of the healing process and memorial events are interactive processes with shared personal experiences, officials' public comments, religious ceremonies and media coverage. Together these expressions take form as a collective remembering.

Immediately after a tragedy, people who are affected commonly express their grief spontaneously by making the disaster and their loss visible in different ways. There is commonly also an increased attendance at religious places by those who do not regularly visit temples or other religious venues (Eyre 2007: 446). A memorial makes the event comprehensible and allows for individual memories. How the memorial is presented and how the event is conducted reflect more the concerns of those who arrange and control the public activities that may surround it. Often, memorials are not created or organized by those who have been directly affected by a memorialized event. In Thailand, most permanent memorials have been made for, rather than by, people impacted by disaster. However, there are also examples of memorials created by the affected people themselves.

Anne Eyre highlights the fact that commemorative activities after a collective tragedy also become an arena for social and political commentaries on a broader change of state and status for a family, community or society. This introduces a self-reflexive way of questioning safety, resilience and vulnerability that they were unaware of before the tragedy (Eyre 2007: 442).

In a devastating catastrophe like the tsunami, whole villages are affected and the healing process is significant for the local community as well as the nation. It is important to note, however, that the official commemoration revealed a social hierarchy that excluded certain minority groups in the local communities from the ceremonies. The way that the official and local ceremonies were organized communicates clearly that the ceremonies served different goals.

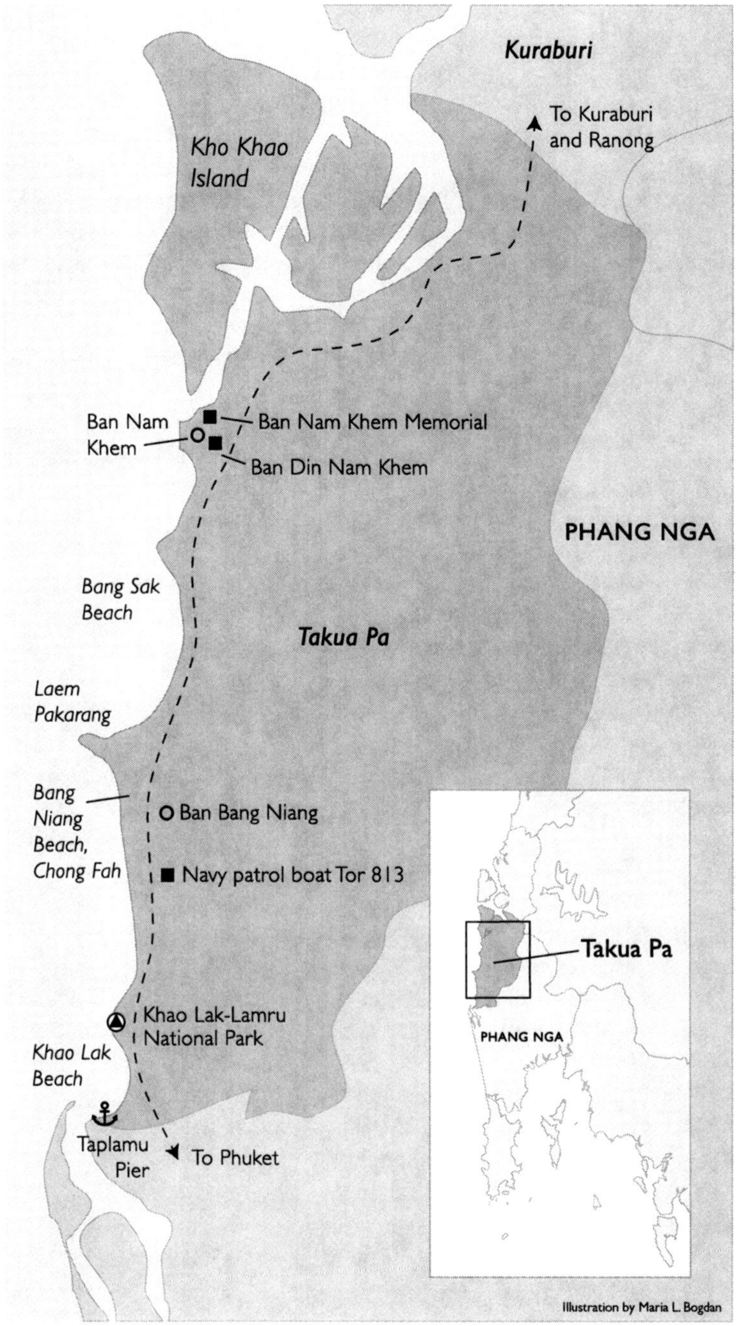

Figure 8.1 Map of memorial sites in Takua Pa (source: illustration by Maria L. Bogdan).

This chapter concentrates both on the memorial practices and on the sentiment that the memorial commemorations and permanent memorials produced among those affected. The official memorials and annual ceremonies that are addressed in this chapter were created to meet the requirements of people from the many nations that were affected by the tsunami. The five consecutive annual memorial ceremonies that followed the tsunami are described and dealt with in relation to the survivors' experiences of tsunami memorial ceremonies. This chapter explores some of the divergences between the memorial ceremonies organized locally by the impacted communities and the memorial ceremonies organized by the Thai government (see Figure 8.1 for locations of the memorial sites of Takua Pha).

Local memorial ceremonies

Photographs of missing and deceased family members were of huge importance to the survivors, and they carried these photos to the temples and ceremonies to honour those who had died (see Chapter 5). At ceremonies, the number of photographs of deceased and missing persons was astonishing. When the first year after the tsunami had passed, the number of photographs brought to the temples decreased and, instead, more people held ceremonies for the deceased privately at home or at the beach where the body had been found.

According to popular beliefs in Thailand, there is, as we have seen, a liminal period after death and most people have the opinion that the spirit is still around for some time after passing away. Services are usually held in the evening: the first three days after the death and then again on the seventh day, fiftieth day, one hundredth day and one-year anniversary. Central in the one hundreth-day ceremony are blessings to all living beings and merit-making for the dead to increase the possibility that they will have as good a rebirth as possible and will be released from suffering.[1]

The one hundredth day is marked by a ceremony known as the "cessation of weeping". Subsequent memorial services can be held on the first three anniversaries of death, the last of which is known as the "termination of mourning". At that point, the obligations of the survivors to the deceased are complete; from that moment on, the deceased is to be treated as an ancestor rather than simply as a person who has died.

At many of the tsunami-hit areas, local authorities organized a one hundredth-day memorial ceremony. In the provinces Phuket and Phang Nga, the ceremonies were of an interfaith character and involved chanting, prayers and short speeches. The memorial ceremony was held in the morning before lunch, outdoors and close to the beach. The ceremony was led by local representatives and leaders from different religions performed the service. I observed that the interfaith memorial ceremonies were not as well attended as the one hundredth-day ceremonies held at the temples.

In connection with the one hundredth-day ceremony, one temple in Phang Nga province organized a special ordination of 100 boys in order to transfer religious merit to those who had perished in the tsunami. Another aim was to also give the boys a chance to talk about the tragedy and to understand and handle their experiences of the catastrophe. The one hundredth-day ceremony arranged by the temples featured Buddhist chanting and Buddhist talks by the monks. Laypeople brought food to the temple and after the monks had finished their meal, the laypeople ate lunch together. During the lunch people had the chance to chat with each other. At all of the ceremonies, the participants transferred merit to those who had perished. This was done through chanting and pouring water from one vessel to another, in the same way as I have described in earlier chapters.

At the temples I visited, some people that gathered seemed to want to talk about what they had been through since the tsunami happened and they spontaneously described their loss of people dear to them. Safety was a theme that frequently came up in conversations and people were anxious and did not trust the warning systems; this fear was accentuated after a new earthquake occurred shortly before the one hundredth-day ceremony.

The official schedule of events for the one hundredth-day tsunami memorial in Phang Nga was held from Saturday to Monday (2–4 April) 2005. The main venues in Phang Nga were at Ban Bang Niang Beach and the Tsunami Volunteer Centre and in front of the Navy Patrol Boat Tor 813. Free transport with local buses (*song taew*) was provided between the venues. The memorial included various activities, with performance of local folk music, dance and boat repairing among many other things. The programme began in the morning and ended in the evening. Interfaith ceremonies were included in the programme, and a Moken ceremony, *Loi Rua*, was performed, along with a service for the Thai king's grandson, Khun Poom Jensen, who died in the catastrophe.

Local annual ceremony

All temples in the tsunami-hit areas hold annual memorial ceremonies to commemorate the deceased. The ceremonies' main objective is to transfer merit to those who have perished. The ceremonies usually begin at 9 a.m. in the morning and end about 12 noon. Some of the temples I visited also invited well-known monks to speak in the afternoon. I participated in numerous local ceremonies and some were more elaborate than others. However, the ceremonies follow the same structure and I will here give a description of the second memorial ceremony at Wat Bang Muang. It was announced that the ceremony was to be held in the morning on 26 December 2006. When my assistant and I got to the temple at about 8.30 a.m. many laypeople had already arrived. Most of them were dressed in white. They had brought containers (*pinto*) with food that they had prepared for the monks. Together with the food, they also brought photos of deceased family members to whom they were going to transfer the merits earned by donations and participation in the ceremony.

The temple's yard was organized to accommodate a rather large gathering. On the ground in front of the main Buddha statue, red and yellow mats were laid out to sit on and white plastic chairs were placed in rows for those who preferred not to sit on the ground. From the main Buddha statue, nine white threads were stretched to a large jar filled with blessed water. One big yellow candle was placed in the middle of the jar and eight small yellow candles were fastened on the brim of the jar in such a way that the melted wax from the candles would drip into the water. White threads were also attached to the place in the temple where the ashes were kept and then the threads went from that place to the displayed photos and further on to the monks' hands. During the ceremony the monks held the threads in their folded hands while they were chanting.

Tables were positioned in the temple's yard for different purposes. Bowls were placed on some tables for the attendees to offer rice. A large table covered with blue fabric was for the laypeople's *pinto*s, together with cooked food that should be offered to the monks during the ceremony. At one table there was a pile of white sheets for people to write the names of their deceased family members and loved ones. Beside the pile was a large tray on which the sheets with the names would be placed to be burnt later. That table was positioned in the same area where laypeople lit incense in respect to the dead. Since the tsunami hit, people had brought photos of their deceased family members and relatives to the ceremonies at the temples. The display table for the photos was the largest table at the two year anniversary ceremony. The table was decorated with black and white fabrics, and many of the photos on the table depicted children. After the tsunami, Wat Bang Muang was used as a makeshift morgue. To remember the difficult period after the tsunami, the temple had erected two rather large boards with photos that showed the situation at the temple.

Shaded by trees, 25 monks sat on the stage. Before the ceremony began, firecrackers were lit three times. This was done to invite the celestial beings (*thewada*) to participate in the ceremony in order to protect the participants. Firecrackers were also set off during the service before the donation of the robes to the monks. It was explained to me that the firecrackers at that time were fired as a signal to the dead to receive the merit (*bun*) that the relatives dedicated to them. After the donation of the robes, the head monk lit the eight candles that were fastened on to the jar and the big candle in the middle. The monks then performed chants to transfer merits to the deceased (*suad bangsukul*) and the laypeople were asked to go to the table with the *pinto*s and donate the food that they had brought to the monks (*sai bat*). People put rice in the monks' bowls, that were placed on a table, and gave the *pinto*s to the monks. After that followed moments of silence in remembrance of those who had died and people concentrated their thoughts on those who had perished.

The merit earned from the robes offered to the monks was symbolically transferred to the dead people during the ceremony. The set of robes donated for somebody who has died (*pha bangsukul*) consists of three pieces. The robes are offered to the monks in order to pass them to those who have lost their lives. At the end of the ceremony, water was used as a medium of transferring merit. Following earlier descriptions, the laypeople poured water from one container to another. A lit candle was placed in the receiving container. The people were joined together during the pouring of water by holding on to each other and the monks chanted while holding the white thread. Water is considered to be the medium that connects the dead with their family and loved ones who perform the ceremony. The light from the candle that was placed in the container was considered to function as a guide for those who had died. The water from all the small vessels was collected in order to be mixed with the ashes from the burnt sheets with names of the dead, and the water with the ashes was then poured beneath a tree. A layperson explained to me that the people believe that donating a full set of monk's robes is similar to being ordained.

During the ceremony, before the donation of the robes, Phramaha Sa Nae from the Wat Tha-Rea Temple in Phuket province gave a speech. His talk contained fundamental Buddhist teachings about life, death and the effects of the tsunami. He talked about how the world is constituted and applied the Buddhist teaching to laypeople's everyday lives. Phramaha Sa Nae listed the objective of the memorial service. He concluded that the ceremony's main aim was to transfer merit to those who have died. Other reasons for performing the memorial ceremony, according to him, were: to share a common fate with people who had lost a family member, support and encourage each other after having lost people who were close to us, express gratitude, remember parents' advice and continue, or begin to follow, the advice they had given when they were alive. The monk said that keeping the tradition of having memorial ceremonies for dead people also maintains Buddhism. He ended in a positive way by encouraging people to look at life's many possibilities.

Several hundred people attended the service that morning and firecrackers were fired to mark the end of the ceremony in Wat Bang Muang. In the afternoon, Myanmar monks and survivors held a Buddhist memorial ceremony at the temple.

The monks had finished their meal before 12 noon. Most people who had brought food to the temple collected their *pinto* from the platform close to the monks and found a place in the temple area to have lunch together.

National annual memorial ceremonies

Memorial services provide an opportunity for those affected to share their grief with others. Official memorial services often have an important national as well as local role and are likely to receive extensive media coverage. It is important to consider the organization and structure of the programme carefully. Important aspects are timing, invitations, representation and conduct. Decisions made about issues such as where dignitaries sit in relation to the bereaved and survivors have symbolic meaning and political consequences. The presence or lack of representation by figures such as senior politicians or members of the royal family might potentially cause resentment and further distress because of the significance attached to participation. This is especially so given the charged emotional atmosphere of such events.

At the same time, where numbers of attendees are restricted, extended family members and survivors can often be left feeling aggrieved and excluded, particularly when the invitation of dignitaries is prioritized over direct victims who are left to observe proceedings from the streets outside or on television. It is understandable that they can be left asking who and what the commemoration is really for.

The first-year tsunami memorial (2005)

The first anniversary after a disaster is often particularly significant, well attended and marked by the mass media and the wider community. In Thailand, one year after the tsunami, a series of memorial services took place at key sites including the temples. The first and subsequent anniversaries of the disaster received much national and international coverage.

The first anniversary of the tsunami occurred on 26 December 2005, and the Thai government invited approximately 14,000 injured people and relatives of those killed in the disaster to attend the ceremonies. Among them were 7,500 Thais and 6,500 foreigners. The Thai Prime Minister, Thaksin Shinawatra, and several Thai ministers participated in the memorial. The Thai Princess, Ubolratana, the eldest daughter of His Majesty the King, had lost her son, Khun Poom Jensen, in the tsunami and she participated in the evening ceremony together with her two daughters. More than 10,000 Thai and foreign visitors, plus 550 international media and 650 local media representatives attended the services. The ceremonies were marked by religious readings, performed by Buddhist, Christian, Hindu and Islamic leaders, the laying of wreaths, a minute of silence and the dedication of a specially commissioned memorial to ensure that future generations never forget what happened on that fateful morning.

The visitors to Phuket, Phang Nga and Krabi provinces could choose to attend tsunami memorial services at Patong Beach, Kamala Beach, Mai Khao Memorial Wall in Phuket province, Bang Niang and Nam Khem villages in Phang Nga province and Phi Phi Island in Krabi. On 26 December, the Thai government held three national events to commemorate the occasion. The first ceremony was held in the morning and took place simultaneously in the tsunami-hit provinces of Phuket, Phang Nga, Krabi, Trang, Ranong and Satun. During an afternoon ceremony held in Phang Nga province, Prime Minister Thaksin Shinawatra laid the foundation stone for the construction of the official Tsunami Memorial on Hat Lek beach in Lamru National Park. In Phang Nga the evening ceremony took place at 6p.m. on Chong Fa beach, with interfaith memorial services conducted by representatives of Buddhism, Christianity, Islam, Sikhism and Hinduism. At this event, Princess Ubolratana delivered a speech on behalf of the bereaved and Prime Minister Thaksin Shinawtra also gave a speech. Two young tsunami survivors, a Thai boy called Patiwat Kongkla and a British girl named Tilly Smith read poems in commemoration of the event.

Shortly before sunset on Chong Fa beach, over 10,000 participants took part in the evening ceremony of remembrance for the victims and the commemorative event highlighting the government's "One Year in Memory of Tsunami Ceremony".

Schedule for the "One Year in Memory of Tsunami Ceremony"

09.00	Tsunami Victims Memorial Service at Phang Nga – Ban Nam Khem and Bang Niang Beach; Phuket – Kamala Beach, Patong Beach and Mai Khao Memorial Wall; Krabi – Phi Phi Islands.
10.00	Speech by deputy prime ministers or ministers as representative of the Royal Thai government.
10.15	Laying of flowers by VIP guests, followed by one minute of silence.
10.20	Laying of flowers by relatives of victims and participants. The ceremony ends at 11.00. The lunch is hosted by the Royal Thai government.
14.00	Departure for Had Lek Beach (for victims' next of kin only). Departure for Chang Fah Beach (for injured only). Tsunami Memorial Foundation Stone Laying Ceremony, Had Lek Beach, PhangNga (for next of kin, VIP guests and diplomatic corps only).
16.20	Arrival of Thaksin Shinawatra, Prime Minister of Thailand.
16.39	Laying of the Tsunami Memorial Foundation Stone.
17.09	End of ceremony. Tsunami Memorial Designs exhibition, poem recital and musical performances.
17.30	Departure of participants for the Interfaith Memorial Services at Chong Fah (Bang Niang) Beach, Phang Nga. Interfaith Memorial Services, Chong Fah (Bang Niang) Beach, Phang Nga.
18.30	Arrival at Chong Fah (Bang Niang) Beach, Phang Nga. Distribution of floating lanterns to participants.
18.30–19.00	Interfaith Memorial Services, approximately five minutes for each religion: Buddhism, Christianity (Roman Catholic and Protestant), Islam, Sikhism, Hinduism and Judaism.
19.00	Arrival of Princess Ubolratana. Video of Tsunami Victims Memorial Service and Tsunami Memorial Foundation Stone Laying Ceremony.
19.10	Report by deputy prime minister. Poem recital.
19.10	Speech by Thaksin Shinawatra, Prime Minister of Thailand.
19.15	Speech by Princess Ubolratana, representative of a tsunami victim's next of kin.
19.20	Minute of silence.
19.21	Musical performance "The Wind of Hope". Lighting of candles and floating lanterns.
22.00	Dinner hosted by Royal Thai government.

The dress code for 26 December 2005: Tsunami T-shirt, slacks/long trousers, delegate badge (T-shirt and badge provided in the delegate kit).

I attended the first-year memorial day in Nam Khem, Baan Niang and Chong Fah, and the Tsunami Memorial Foundation Stone Laying Ceremony at Had Lek Beach. The survivors from foreign countries, families of the foreign victims and others who attended the memorial services praised the government for organizing such a moving service.

However, there were also complaints about the lavish cost of the first-year memorial. The majority of the returning survivors of the disaster paid for their own journeys back to the tsunami-hit regions. For some, coming back to the tsunami area served as a form of therapy and helped them to deal with the trauma they suffered. Many of the mourners preferred to grieve solemnly in private. Several foreign nationals held their own memorial services at the sites where their compatriots died, independent of the official ceremony hosted by the Thai government.

The second-year memorial (2006)

I attended the second-year memorial ceremonies that were held on 26 December in Phang Nga. This time, they were not as grand as the first year memorial and were not arranged by the Thai government. People in Phang Nga and the other tsunami-stricken areas gathered on beaches and in villages and held commemoration ceremonies to mark the tsunami that pounded the Andaman Coast two years ago. Similar to the first-year services, the memorial services were held at the Ban Nam Khem tsunami memorial and in front of where the navy patrol boat Tor 813 was washed ashore on Khao Lak at Ban Bang Niang. In Krabi's Muang district, about 400 people – government officials, locals and foreigners – joined in Buddhist, Christian, Islamic and Chinese rites on Ton Sai beach at Koh Phi Phi.

At the tourist destination of Khao Lak, about 300 Thai and foreign survivors joined a ceremony to pay tribute to the victims, observing a minute of silence while Thai police placed flowers and incense on a boat that was washed ashore by the massive waves. Some people laid white roses on the beach to honour the victims, some wept as they looked out across the sea that brought tragedy, death and destruction two years previously.

At the second-year memorial, Princess Ubolratana Rajakanya attended a night ceremony at Bang Niang Beach, Chong Fah, where her son, Khun Poom Jensen, lost his life. Candles were lit and poems of condolence were read by victims' relatives and survivors.

The ceremony in Takua Pa district in Phang Nga held one of many low-key religious events organized by local communities and with support from the government. The ceremonies were generally less traumatic than the first annual ceremony, but for many the pain remained sharp. The main ceremony in Ban Nam Khem was chaired by a representative from the Prime Minister's office. In Ban Nam Khem, many of the villagers had experienced a difficult year, with far less support than they had expected. When the minister greeted the people who had gathered, the villagers took the opportunity to complain to the him that the promised relief did not reach them and they described how they had been ignored by the government in the difficult time following the tsunami.

In Phuket, the Tourism Authority of Thailand, The Phuket Provincial Administration Office and Patong Municipality jointly hosted the ceremonies on 26 December to remember and honour those who died in the tsunami. The ceremonies began at 9.30 a.m. and at 10.00 a.m., nine Buddhist monks performed a merit-making and blessing ritual. Participants made merit-making offerings to the monks followed by the Buddhist monks' blessings. The ceremony ended at 7 p.m. with a wreath-laying and candle-lighting ceremony and traditional *khome loi* floating lanterns were released in memory of the deceased.

Also in Phuket, both Thai and foreign people attended a special memorial ceremony in which 99 monks chanted for the victims and a great number of flowers were placed on the sand. At this ceremony, the local officials informed the participants about security issues including a tsunami warning system to prevent future disasters. It was reported by the local news that the service was mostly conducted in silence, with only the sounds of the monks praying and the waves on the beach heard.

The Tsunami memorial service at Kamala in Phuket province were organized by the Phuket Japanese Association. It was held at the Andaman Tsunami Victims Monument at Kamala Beach on 26 December in the afternoon and ended at 4 p.m. with a Japanese Buddhist blessing ceremony. The memorial service at Kamala was held in the afternoon. The President of the Phuket Japanese Association, Phuket's Governor, the Japanese Chamber of Commerce in Bangkok and a representative from the World Fellowship of Buddhists held speeches during the ceremony.

In the evening, more than 1,000 people gathered at Loma Park in Phuket for a commemoration ceremony. It was opened with a speech by the Tourism Minister. Local students performed songs in memory of those lost to the tsunami waves. There was a one-minute silence, followed by the laying of wreaths on the beach. Twenty thousand candles were lit and placed along one of the beaches, 5,000 floating lanterns were then released into the night sky and white roses were distributed to commemorate the victims.

The third-year memorial (2007)

I attended the third annual ceremony in Phang Nga province and it was even more local than the second annual service. In Khao Lak, a small morning memorial ceremony took place at 9 a.m. at Baan Bang Niang. Speeches were held and flower offerings were carried out in remembrance of the tsunami victims. The ceremony marked the formal recognition of the patrol boat Tor 813 as an official tsunami memorial.

The memorial in Nam Khem was announced as: "Three years after the Tsunami: Come together for the recovery of the Andaman". The memorial was held at Ban Nam Khem Tsunami Monument with a multi-faith religious ceremony that involved Buddhists, Muslims and Christians. This year, the Moken was included in the programme and they conducted a floating boat ceremony as part of the proceedings. The main part of the programme was devoted to community discussions on topics related to lessons learnt and problems associated with recovery after the tsunami.

The programme was divided in to morning, afternoon and evening sessions.

The memorial programme

- Morning session:

  08.00–11.00 – "Tsunami Walk Rally" from Ban Nam Khem School to Ban Nam Khem Memorial, and from Ban Nam Khem Temple to Ban Nam Khem Memorial;
  traditional ceremony;
  making merit for the people who perished in the tsunami;
  grand merit ceremony for the community;
  floating raft ceremony to banish bad luck;
  Planting mangrove to create natural wave protection;
  gathering of more than 200 local fishing boats.

- Afternoon session:

  brainstorming in discussion forum: "Experience of the post-recovery in the community";
  local sports competition.

- Evening session:

  local cultural performance to commemorate people who passed away during the Tsunami with a floating lamp ceremony;
  exhibitions addressing several issues:
  land conflict after the tsunami.
  communal and ecological matters.
  the effects of implementing mega-projects in the communities;
  migrant workers;
  gender issues;
  increasing the capacity to be prepared for post-disaster situations.

  The dress code for the day was yellow dress in veneration to the Thai king.

On 26 December at Bang Maruan, Takua Pa and the Thai Tsunami Victim Identification Centre and Cemetery, ceremonies began at 1 p.m. with a special service conducted by representatives of Buddhism, Islam, Christianity and, finally, Sikhism. This was also a cemetery and here were still almost 500 unclaimed bodies, which had been kept in stainless steel containers 1.5 m underground. The graves are marked with letters and numbers.

Some of the Buddhist temples in the area offered Burmese monks the opportunity to hold ceremonies for Burmese inhabitants in the area. On 25 December, local Burmese people had gathered for a memorial and to mourn at Wat Bang Muang.

HRH Princess Ubolratana presided, as she had done at every anniversary, over an evening ceremony at Chong Fah beach in Bang Niang. Twenty-two foreign ambassadors, diplomatic members and representatives from two international organizations attended the memorial ceremony. The evening ceremony began at 6 p.m. with representatives of the five religions mentioned above.

At 7 p.m. Princess Ubolratana arrived with her daughter Khun Sirikittiya to lead the ceremony. The princess and her daughter visited the exhibition to remember Khun Poom Jensen. It showed photos of Khun Poom's life and his home. A special project for the education and development of children, run by Wahira Hospital in Phuket, was presented at the exhibition. Also on display were examples of Phang Nga province's preparedness to deal with emergencies.

A local student who lost her father in the tsunami read a poem in Thai and the princess gave a speech before lighting a candle and observing one minute of silence. After that, the princess released the first of 5,395 white lanterns (one for each victim) up in to the sky, as a tribute to those who lost their lives.

The fourth-year memorial (2008)

In Phuket, the Governor, Khun Preecha Ruangjan, led the participants in a session of reflection and observed a one-minute silence in respect to the victims. Survivors from other countries participated and shared their experiences. Buddhist monks chanted and school children played traditional Thai instruments.

At the Mai Khao Wall of Remembrance, a simple multi-faith religious and wreath-laying ceremony for victims from 46 countries took place in the morning. One minute of silence was observed and the Phuket Vice Governor led the participants in laying wreaths and flowers at the wall. In the morning, a religious ceremony and commemorative event was held in a public park in Kamala with locals and visitors attending. They also offered alms to Buddhist monks. In Phuket, the memorial day ended with the "Light Up Phuket" ceremony, which was held on a stretch of Patong Beach.

At the Bang Muruan Thai Tsunami Victim Identification Centre and Cemetery in Takua Pa, almost 400 unidentified bodies were buried. Commemoration of the fourth anniversary of the tsunami held at the cemetery was a low-key event. The police department held a remembrance ceremony at the cemetery and observed one minute of silence in respect to the victims. Members of the Thai Tsunami Victim Identification Unit placed large white wreaths at the memorial. In 2008, villagers scattered sunflower seeds at the cemetery and they have since grown wonderfully.

At Ban Nam Khem Tsunami Memorial Park, the local Bang Muang authorities and the tsunami victim network conducted joint religious ceremonies (Buddhist, Christian and Muslim). Survivors had placed framed photographs of their loved ones on a large table. Subsequently, the Thai government's Education Minister led villagers who queued one-by-one to leave white roses in memory of those who had perished.

In Bang Niang, the route to the patrol boat Tor 813 has become known as *Soi Rua 813*. At 8.30 a.m. on the memorial day, a ceremony was conducted at the boat. Flowers had been placed at a shrine on the bow and on the boat deck. Thai people and foreigners laid flowers in remembrance of the victims.

Also in this year a ceremony commemorating the life of Khun Poom Jensen was held. Lanterns to honour those who lost their lives were released at 6 p.m.

The fifth-year memorial (2009)

Five years after the tsunami, the day was remembered nationwide. Monks gathered throughout the country's 76 provinces and 500,000 of them commemorated the anniversary.

In Phuket, the fifth memorial day of the tsunami began with a wreath-laying ceremony for families of victims from 46 nations at the Wall of Remembrance in Mai Khao between 8.30 a.m. and 12 noon. The Mayor of Mai Khao made a speech at 8.30 a.m. This was followed by Buddhist, Muslim and Christian services and one minute of silence.

In the morning, at the Tsunami Memorial Park in Kamala, a multi-faith ceremony was held with representatives from Buddhist, Sikh and Muslim faiths. At the same venue, a Christian ceremony was held in the evening.

In Phang Nga, at 6 a.m., about 1,000 Buddhist monks' alms presentations took place in the main street in front of the police station in Takua Pa town. Later the same day, the main locations for the commemorative events in Pang Nga were Nam Khem, Bang Niang at the patrol boat Tor 813 and the cemetery at Bang Maruan.

At the Tsunami Memorial Park in Ban Nam Khem, the interfaith memorial service began at 10 a.m. Photographs of family and friends formed the centrepiece of the occasion. Education Minister, Jurin Laksanavisit, chairman of the ceremony, laid a wreath in front of the picture of Khun Poom Jensen. Afterwards, an interfaith remembrance ceremony was held. Later, the Education Minister presided over the candle-lighting ceremony and the release of lanterns.

At 11 a.m. the police department held a remembrance ceremony at Bang Maruan Cemetery, where 398 unidentified victims are buried. Representatives of 13 embassies observed a wreath-laying ceremony at the cemetery.

The sixth-year memorial (2010)

In Phuket, the sixth tsunami commemoration had its first ceremony at the tsunami Wall of Remembrance at Mai Khao from 8.30 a.m. to 12 noon. Activities included a multi-faith ceremony, one minute of silence and the laying of wreaths. The service at the tsunami monument at Kamala Beach was led by the Phuket Japanese Association from 11.10 a.m. At Loma Park, Patong Beach, there was the "Light Up Phuket" event. Drawing and painting competitions for all ages took place in the afternoon, and there was a traditional Thai musical performance, a recital of a tsunami memorial poem, one minute of silence, religious reflection, the lighting of candles and wreath laying.

The Ban Nam Khem villagers held a memorial ceremony at Ban Nam Khem Tsunami Memorial Park in the same manner as the previous year, with interfaith religious ceremonies forming part of the occasion. The villagers brought photographs of those who had perished in the tsunami and the photos had a central place in the ceremony.

In 2010, the authorities in Phang Nga announced that they had decided that from then onwards the official commemoration of the tsunami would be organized every five years.[2]

The villagers have continued to organize annual memorials, but without prominent participants from the government to conduct the ceremony. Religious representatives have been invited to hold interfaith services and the ceremonies during the daytime follow the same structure as before except for the lack of eminent government participants. In the evening they organize entertainment with music.[3]

Alternative annual memorials

While Thailand and representatives from other countries united to remember the victims of the tsunami at the annual memorial ceremonies described above, some survivors boycotted the events, complaining about the focus placed on the international relations and the neglect of the local survivors. Local memorial events were organized in opposition to the commemoration that involved the government. The local people's memorial organizations invited grassroots groups to join them. They highlighted that the poor lived under severe difficulties in the affected provinces, which still suffered from the impact of the tsunami.

Some survivors revealed that they did not have the strength to join the official memorials because it was too painful for them. They felt that their life was a nightmare, that they were struggling every day with their loss and that it would be too traumatic and emotional to participate in the government's memorial. Some also felt that the government had not done anything to help them.

At the first annual memorial, about 30 of the volunteers gathered at Wat Yan Yao and held a reunion and commemoration. They were among the several hundred volunteers that had worked with the forensic operation at the two temples Wat Yan Yao and Wat Bang Muang and also at the cemetery Bang Maruan, all in Takua Pa district. They had worked together with the forensic expert Porntip Rojanasunan, who was highly esteemed by the survivors for her skill and respectful treatment of all the corpses. The volunteers walked around the temple grounds that had, for 40 days, been their workplace and where they had worked long hours under highly stressful circumstances. One year after the catastrophe, the temple was quiet; the volunteers remembered how crowded the temple was a year ago and they shared experiences and information about the people who were there the previous year.[4]

Some of the Moken and Moklen people who were badly affected by the tsunami complained that they were not included in the official memorial ceremonies. The Moken and Moklen organized their own commemoration the day before the official memorial ceremony. Together with NGOs and other local community groups, the Moken and Moklen held seminars, exhibitions, performances and markets to commemorate the tragedy. They said that they wanted to organize their remembrance on the day before the official ceremonies so people could participate in both memorials if they wanted to do so.[5]

The special ceremony that the Moken held was a boat floating ceremony, with offerings that were set adrift to appease the spirits of the sea. Carved "dolls" in numbers that were equivalent to the Moken family members in each household were put on vessels for the ceremony. Performing the floating ceremony is an old Moken tradition and it is carried out annually. It is believed to help to reduce bad luck and to bring happiness into their lives.

The boat used in the floating ceremony is about five metres long and beautifully decorated with flowers. Before the ceremony begins, the members of the families voluntarily put pieces of their nails, hair and rice into the boat. Then the village chairman performs a ceremony before releasing the vessel into the deep sea with the aim of floating away their unfortunate luck.

Banishing bad luck resembles one aspect of the Thai annual floating ceremony *Loy Kratong*. This ceremony is also a way to predict the coming year and the Moken believe that their future can be decoded by the stability of the boat. If the boat is washed back to the shore it means that they will struggle to earn a living. But if the waves do not wash it back and the boat is out in the sea without capsizing, it means that the Moken's coming year will be fortunate. The Moken inherited the ceremony from their ancestors and they continue to retain the ceremony by holding it once a year between May and December. The floating boat ceremony is also a way to pay tribute to, and have faith in, their ancestors.

The local officials in Phang Nga invited Myanmar labourers to join the first-year memorial service in the province. However, it was an ordinary day for most Myanmar migrant workers and therefore not all were free to join in. Those who participated shared their traumatic experiences and commemorated their friends and colleagues who were killed by the tsunami. Most of the Myanmar migrants who survived the tsunami mourned together with their Myanmar friends. One year after the tsunami, many Myanmar migrants were still struggling to cope with the hardship following their job losses after the tragedy. According to local human rights groups, the Myanmar migrants seemed to have been forgotten and they demanded that the government take better care of them, since they had also lost their loved ones and their possessions in the tsunami (Hulme 2005).

Survivors' reactions

The first official memorial ceremonies received plenty of praise from foreigners and prominent guests, yet they did not attract all local people and many chose instead to hold private services for their close family, relatives and friends who had died. Some local survivors who had lost their loved ones considered it awkward to join the official ceremony, which had the aim of highlighting the progress of the government's effort in the recovery process. I met many in Phang Nga who were dissatisfied with the government's rebuilding efforts and said that they were a long way from completion. For some local people, the reason for attending the official memorial might be to meet some of the cabinet members who they could hopefully speak to and complain about their problems.

A common reaction to the official ceremonies was that too much money was spent on the services while villages remained in ruins. The local villagers thought that the ceremonies were too exclusive, extravagant and commercially driven. Some agreed with the government that the ceremonies would help the tourism industry to promote the image of the nation. Nevertheless, they believed that the ceremonies should be simpler and show true regret for the deaths and for the affected families. Some considered it unnecessary to hold annual ceremonies. They thought that a memorial for relatives of the victims where flowers were laid as commemoration to their loved ones was enough. One reason for this opinion was the concern that if the annual ceremony continued, the people would not be healed from the experiences of the disaster. Many said that the ceremony re-opened wounds. Some survivors did not participate in the official ceremonies because they did not understand the real objectives of the event. They said that they wanted to forget what had happened, but the memorials brought back their sadness. Others said that it was better to go to work rather than attend the ceremonies. There were complaints that the money that was spent on the official memorials could have been put to better use helping the victims. Private donors and foundations had financed the rebuilding of some villages. The villagers that I talked to who lived in these communities told me that they had not received any money from the government.

At the first-year memorial, I also participated in the afternoon's Tsunami Memorial Foundation Stone Laying Ceremony at Khao Lak-Lamru National Park. A teacher who attended the ceremony said that he felt that it was too costly and that it was especially problematic since the villagers' struggles were not yet over. At times, the event resembled a stylish concert at an exclusive beach and it had a festive atmosphere. This teacher said that they were selling tragedy for tourism.

Permanent memorials

The importance of remembering the past is illustrated by the memorial sites and museums that serve as permanent reminders of the catastrophe. Memorials exist in order to remember an event and to honour the victims of the disaster. As public symbols, memorials should be congenial to a heterogeneous variety of individuals with diverse experiences of the catastrophe. In order to encompass the wide range of affected people, public memorials undoubtedly must have a neutral expressionMemorials are commonly located in a park or in other quiet surroundings. Generally, the time and place of the disaster are mentioned, together with how many perished. Some memorials also have names of those killed. Simpson and de Alwis suggest that a memorial compensates for the terrible event that has occurred and they suggest that the fundamental function of a memorial is to moderate survivors' experiences and memories of what happened to them (Simpson and deAlwis 2008). In combination with presentations of experiences and facts, memorials and memory also include, as Simpson and de Alwis have highlighted, a form of politics (ibid.).

The majority of the Thai national tsunami memorials are symbolic and include abstract objects. In contrast, some of the local memorials are real objects from the tsunami that have been kept as evidence and reminders of the tragedy. Examples are certain boats that have been left at the spot where they were washed up by the tsunami waves.

In July 2006, the tsunami memorial in Kamala, Phuket, was officially opened. The memorial forms the centrepiece of a new park in the town and is a ten-metre-high metal sculpture called Jitt Jakawan (Heart of the Universe), designed and built by the Thai sculptor Udon Jiraksa. On 30 September 2006, Thailand's first underwater tsunami memorial was laid on the seabed in Tonsai Bay, off the island Ko Phi Phi Don. This unusual memorial is a project of the Transport Ministry, Thai Airways and the Tourism Authority of Thailand and lies at a depth of 20m. The monument comprises three large granite markers, each being 150cm in height and arranged to form an equilateral triangle, with a smaller 80cm granite marker standing in the centre. The three main pyramid-shaped granites have square bases, and are set in a triangular pattern to represent the elements of earth, wind and water. The triangulation indicates the natural balance between these elements. Each face of the pyramid granites is engraved with memorial statements from the embassies (in Thailand) of countries whose citizens lost their lives in the tsunami. The foundations of the pyramids are engraved with words of wisdom from various philosophers to remind us of the right way of living. The distance between each of the markers is 5,395cm and corresponds to the number of victims whose names and nationalities were identified. Three yellow buoys on the water's surface indicate the position of the granites. The smaller granite stand is located in the middle and has a narrative description of the tsunami event of 26 December 2004; it represents a human who is surrounded by the elements of earth, water and air. In addition, a Thai style pavilion (*sala*) can be seen situated in a northerly direction from this underwater monument. The *sala*, which was built from wood collected from the tsunami wreckage, lies at a distance of 2,874cm from the memorial site. The distance represents the number of victims whose identity could not be determined.[6]

Mountains of remembrance

In 2005, the Thai government held an international tsunami memorial design competition to build a memorial and visitor's centre in Khao Lak Lamru National Park. Designs were submitted from 379 individuals and teams, from 43 countries. The result of the design competition was announced in May 2006. The winners were two Spanish architects, in collaboration with Phuket-based Naga Concepts (Thailand) Co. Ltd. Their design was named "Mountains of Remembrance". The jury made their decision after considering feedback from visitors to the Architect 2006 exhibition in Bangkok in May, and from local people surveyed at Takua Pa District Office in Phang Nga province. The winning design is intended to represent "a temple hidden in forest, light, nature and silence". It will recreate an artificial piece of nature, integrated into the Khao Lak Lamru National Park. The jury wrote that they were impressed by:

> the openness, generosity and ecumenical quality of this scheme which combines references from many different cultures and religions in providing an iconic landmark and place of reflection for the families of victims and survivors of the Tsunami Disasters as well as for the general public.[7]

The scheme combines references from several cultures and religions, and provides an iconic landmark and place of reflection for the families of victims and survivors of the tsunami disaster as well as for the general public. A path through the forest, with meeting points and resting areas, will lead to the memorial site and then to the sea. At the highest point, five metallic glazed towers of differing sizes will be erected in a cluster and interconnected via a covered square: the Memorial Tower (the largest at over 30 m in height and designed for reflection and meditation), the Museum Tower (with a lookout platform), the Warning Centre (for education), the Amphitheatre Tower (for presentations), and the Restaurant and Shop Tower (for relaxation). The project was planned to open in 2008 but the "Mountains of Remembrance" has not been realized. The project has not been granted necessary permission to use part of the National Lamru Park for the memorial. Environmental experts and academics resist the memorial project location and argue that the site selection is very unfortunate and they do not want any construction in a national park because it will harm the environment. Furthermore, the financial side of the project was not yet settled when the memorial project was postponed. However, at the first anniversary of the tsunami disaster, Prime Minister Thaksin Shinawatra had already resided over the foundation stone-laying ceremony at the memorial construction site. The project is still pending and there has been no sign of activity on the project since Thaksin Shinnawatra's cabinet ended in 2006.

Ban Nam Khem Memorial Park

On 25 December 2005, the construction of the Tsunami Memorial Park in Nam Khem was completed. The memorial site covers two acres and is located close to the sea at Laem Son Beach in Takua Pa district. The project was initiated by an army chief who cooperated with local and international organizations to raise money for the project. A German volunteer foundation donated almost 15 million baht and the Royal Thai Army contributed seven million baht. A special memorial walkway was built alongside two walls and one of the walls has the names of some of the people who perished in the disaster. In Phang Nga, more than 400 large and small fishing boats were lost in the tsunami and some boats were carried inland by the waves. The walls alongside the memory lane, opposite the wall with the name plaques, are made of polished black stone and built around one of these boats. The 15 metre-long tidal wave walls are located at the entrance to the park and the walls symbolize walls of water. In the middle of one of the walls, a fishing trawler has been left on the spot where it was washed up by the tsunami waves.

In the park there is an open space for commemorative events, an open pavilion and two buildings, one for showing places of interest and the other for selling local products. Later, one more building was added which serves as a small tsunami museum.

A statue of Kromluang Chumphon Khet Udomsak, the founding father of the Thai Royal Navy, was erected in the park. He was the son of King Chulalongkorn, Rama V, and he is also called Prince Abhakom of Chumphon. He was educated at the Naval Academy in the United Kingdom and was Commander-in-Chief of the Thai Royal Navy. He contributed to the modernization of the Thai Navy and is revered by mariners and Thai people. He passed away in 1923 when he was 44 years old. A Chinese shrine has been placed on the grounds of the memorial and villagers make offerings both to the statue of Prince Abhakom of Chumphon and to the Chinese shrine before going fishing. A Buddha statue was later placed on Nam Khem's shore facing inlands.

Tsunami memorial boats

The most well known of what I would call the "tsunami memorial boats" is the Navy Patrol Boat Tor 813. That boat was carried about 2 km inland and remains in Bang Niang, Khao Lak, where it came to rest after the tsunami. The boat was on patrol, guarding the king's grandson as he jet-skied in front of La Flora Resort where he was on holiday. The king's grandson died in the tsunami and the other members of the royal family who were also at the resort were saved by taking refuge on the higher levels of the resort's buildings.

Three memorial boats are left inland in Nam Khem. These 60 ton vessels were swept ashore by the tsunami waves and are now true evidence of the how powerful the tidal waves were. One trawler ended up in what is now Nam Khem's memorial park and is a part of the memory lane. The two trawlers named "Blue Angel" and "the Demon" (*Sri Samat*) were, for a long time, left where the tsunami had deposited them. The story about the "Blue Angel" describes how the boat was swept through the village by the waves, amazingly through relatively open ground, and caused no deaths. In contrast, the vessel "Demon" demolished many houses and caused many deaths on its way through the village. The two boats were not at sea when the tsunami hit; they were moored at piers in the village. The "Blue Angel" was at Por Songpoi pier and had several passengers aboard. A member of the crew saw the tsunami wave coming and tried to reach open water, without success. The story also mentions that when the "Blue Angel" swept past a father and his young son, the father managed to grab onto the boat and thus both the father and his child were rescued by it. The "Demon" was docked at Pho Noi pier when the tsunami reached Nam Khem and the second wave carried it into a crowded market road with disastrous consequences. As a result of land issues, both trawlers have subsequently been repositioned on some open ground not far from where they were washed up and villagers in Ban Nam Khem said that moving the "Blue Angel" went smoothly but moving the "Demon" was costly and troublesome and the whole village had no electricity for almost a day because they had to cut the electrical cables in order to move the vessel.

Anne Eyre points out that it is increasingly recognized that the survivors, family and relatives are key stakeholders who need to be consulted in the planning process of permanent memorials (Eyre 2007: 453). In Thailand individual survivors were not involved in the work with memorials built by the government or those that were sponsored by big enterprises. As already mentioned, some of my informants said that they did not feel included in the "national" ceremonies. They said that the "national" ceremonies were only for some Thais and for the foreigners that died in the tsunami. An example they gave was that the name of their family members and relatives who had perished in the tsunami were not mentioned in the memory lane in Ban Nam Khem and once again, they felt ignored.

Memorial museum and learning centre – Ban Din Nam Khem

Since the tsunami, a number of commemorative sites have been established by those directly affected. Survivors who took interest in a local initiative to build a permanent memorial site stressed that the memorial should be simple and natural and no one should be excluded. It was important that all the names of those who had died in the tsunami were mentioned. I went with two of my informants to the local tsunami museum, which was part of a cultural and learning centre called *Ban Din* (house made from soil). The museum is located in one of a number of houses built of red soil in an environmentally-friendly manner. Some of the houses are used for accommodation and some have seminar facilities. There are plans for the centre to be used for workshops and training programmes for groups who stay for a couple of days. The museum is divided into five rooms. The museum shows, through text and pictures, the background of the locality. The fishermen's lives in Ban Nam Khem are displayed alongside the history of the rise of the mining industry in the village. Artists belonging to the Andaman group have created the paintings of Nam Khem's history and culture. In this local museum the tsunami is explained in the context of the village. There are also explanations of what a tsunami is, how it develops at sea and how people could survive the disaster. The story about the two boats, "Blue Angel" and "Sri Samat", mentioned above, is on display and the two vessels are also depicted.

One room that I found very special was small and cylindrical in form, without any windows. On the round wall, from the floor to the ceiling were all the names of those who had perished in Ban Nam Khem neatly written by hand in black ink on the reddish clay. The names were written in small letters and it was overwhelming standing in the small room surrounded by the names of all the dead villagers. One of my informants, who had lost family members in the tsunami, went into the room and showed me where to find her family members' names by pointing at them and then she left the room quickly without saying a word. When I came out she explained that it was too painful for her to be inside the room. She also said that this memorial was executed in the right way and she was content to see that her relatives were recognized there. Her dead family members' names were missing on the wall of the memory lane. The memory park in Nam Khem was partly financed by a German company and many German names of those who died in the tsunami appear on the wall of the memory lane. The memory park is located in what was the hardest-hit village in the country and it is apparent that the organizers' aim has not been to include all the names of the villagers who died in the tsunami; it is unclear what the criteria were for choosing the names.

An incident that explains how hurtful it might be to be ignored was described by one of my informants. On the opening day of the memorial park, she said she met a Western woman who gave her a rose. At first, that made her glad and she thought that the rose was a gift for her to take home and put in water. When she was later told that she was supposed to put it on the wall of the memory lane, she also thought that this was fine because she had lost so many loved ones in the tsunami. But when she walked through the lane with the rose in her hand, looking for the names of her close friends, her mother, relatives and others she had lost, she realized that their names were missing and she felt traumatized by the Western woman who gave her the rose. She lives in a rebuilt house not far from the park and she said that she would never walk along that lane again.[8]

Concluding remarks

The tsunami became both a national and international catastrophe, and people all over Thailand and in other countries mourned together with those who were directly affected by the waves. The increase in global travel has given disasters an international dimension and most disasters involve people from more than one country. People from different cultures and religions come together in grief and mourning rituals. Media coverage of disasters has contributed to the level of participation in acts of remembrance and commemoration. In the tsunami, the international aspect was highly recognized by the government, to the extent that the local survivors were set aside.

In a devastating catastrophe like the this, the healing process is important for the local community as well as the nation. The distinctions between the memorial ceremonies that were created locally in Phang Nga and those that were formed by the government lay in both the significance paid to religion and in the politicization of the ceremonies. The national annual memorial ceremonies were created without precedent. They were unique acts that over the following years became more modest and somewhat routinized. The local ceremonies followed the traditional

Buddhist way of conducting memorial ceremonies and what was novel was the size of the number of people that were directly affected by the disasters.

The material presented in this chapter demonstrates that memorials, as manifestations of communal grief, are sensitive in many ways. The survivors' experiences of the different tsunami memorials have been both traumatizing and healing depending on involvement and the character of the ceremonies. The different memorial ceremonies had different emphases and aims, which were not communicated clearly to the survivors. Among the survivors, not only the memorial ceremonies but also the official memory site evoked feelings of unjust treatment. A memorial service and memorial site should be appropriate for those who are involved in creating the memorials for the departed and also for the survivors. The exclusion and neglect of survivors in the official memorial services and their lack of representation in the permanent memorials risked fostering additional trauma.

There seem to be some common aims and functions that can be observed in post-disaster ceremonies and acts of remembrance. Memorials are an important part of grief and mourning, helping to mark transitions of time and status for individuals and communities. Ritually commemorating disasters through the observation of periods of planned silence appears to be part of official memorials worldwide. The post-disaster official memorials after the tsunami in Thailand addressed a heterogeneous setting of people and were constructed as a mix of new and well-known elements, and interfaith religious ceremonies also played a part. The memorial serves many purposes and one important aim is to make the catastrophic event comprehensible.

Notes

1 Information from monks and laypeople.
2 See www.khaolak.net/khaolak/tsunami-commemoration.html (accessed on 4 June 2014).
3 Information from monks and survivors.
4 Interview at Wat Yan Yao January 2006.
5 Personal communication.
6 See www.kophuket.com/phuket/homemenu/tsunami-commemoration.html (accessed on 4 June 2014).
7 See www.okmd.or.th/en/knowledge_detail.asp?id=179 (accessed in July 2011).
8 Interview, July 2007.

References

Eyre, Anne (2007) "Remembering: Community Commemoration After Disaster". In R. Havidán, E. Quarantelli and R. Dynes (eds) *Handbook of Disaster Research.* New York: Springer Press, pp. 441–455.

Hulme, John (2005) "Forgotten tsunami victims: Burmese immigrants in Thailand". World Socialist Web Site: News & Analysis: Asia. Available at: www1.wsws.org/articles/2005/feb2005/burm-f02.shtml (accessed 4 June 2014).

Simpson, Edward and Malathi deAlwis (2008) "Remembering Natural Disaster: Politics and Culture of Memorials in Gujarat and Sri Lanka". *Anthropology Today* 24 (4): 6–12.

9 Building resilience

The tsunami brought heavy losses in terms of lives, means of livelihood, homes and material possessions. In Thailand the magnitude of the disaster, the length of time before corpses were found, experiences of ambiguous loss, the unpreparedness for a disaster and the uncertain livelihood situation after the tsunami prolonged the recovery phase for many survivors. This chapter addresses factors that are considered to be important for coping and resilience building, such as social capital, social networks, ceremonies, rituals, means of livelihood, early warning systems, spiritual faith and individuals' control over their situation. Here the importance of religion for resilience building in post-disaster situations is a main point. The chapter gives examples of self-help groups that emerged after the tsunami and it contains survivors' personal experiences of being members of these groups. Many of the members of the groups had lost the social support networks they had before the tsunami and had to create new belongings, which the self-help groups facilitated.

The development of the concept of resilience was briefly dealt with in Chapter 2. Resilience building may be carried out in different dimensions, such as: personal and familial; organizational and institutional; economic and commercial; infrastructural; and governmental and public safety (McCreight 2010: 4–5). Aldrich (2012) defines communal resilience as the ability of a neighbourhood to engage in positive, networked adaptation after a crisis. He finds that communal resilience is very much about the capacity of the neighbourhood and how effectivly it can cooperate during the recovery from a disaster (Aldrich 2012: 7). Aldrich's definition is useful in understanding the capacity in building resilience collectively. A definition of human resilience that is frequently used is the ability to bounce back to a level of functioning equal to what it was before the disaster (Norris *et al.* 2008: 127). After the tsunami, I found hardly any survivors in Phang Nga who had successfully "bounced back" and lived as they did before the catastrophe. The survivors needed time to recover and rebuild their lives. For some, recovery was faster than for others, and for those who experienced ambiguous loss, recovery was slower.

After a disaster has struck, it is the immediate circle of friends, neighbours and the local community that provides assistance. When nearby resources are scarce or difficult to access, it is those people who have (and use) networks and connections outside the local community that are likely to be better off (Aldrich 2012: 46). The Buddhist *sangha*'s wide network and high level of social capital made it possible to mobilize quickly and at short notice, monks from other parts of the country who could do so, travelled and assisted the temples in the affected areas. A downside is the lack of gender equality in the *sangha*, which has a negative effect on the *mae chi*s' social capital, and limited possibilities for housing *mae chi*s at the temples in the disaster-hit areas. That had serious consequences for the survivors who, only to a very limited extent, received aid and assistance from female renunciants.

Rebuilding the physical structures is, of course, of great necessity but building inner strength and human resilience is similarly important and sometimes of even greater significance. Furthermore, networks have been shown to be more sustainable than buildings (Olshansky 2006: 17; Aldrich 2012: 166). Research in the psychology of spirituality has revealed a wide range of responses to the inquiry of whether faith is helpful in the recovery after a disaster. Some said they experienced that they were protected and taken care of by God or another deity. Others said that faith did not make any difference, and still others suffered because they viewed the disaster as an expression of God's wrath (Brenner *et al.* 2010: 84–85; Pargament *et al.* 2000). Research has also shown that religious coping strategies resulted in better mental and spiritual health among those who used religion in the coping process than among those who did not employ religion in this way (Boss 2006: 84–85; Hill and Pargament 2003).

This chapter discusses different ways of building resilience. It begins by addressing resilience and ambiguous loss and the importance of finding meaning in unforeseen deaths – answers that are sought after in religion. Examples are given of how religion, social capital and resilience building are interconnected. The importance of a Buddhist reform group's relief work and the activities of self-help groups that emerged after the tsunami is explored in this chapter. Being able to support oneself is important for recovery. Many survivors had difficulties finding a livelihood after the tsunami and they even tried supernatural ways of dealing with misfortune, which this chapter covers. Resilience building and a high level of social capital require trust (Aldrich 2012: 40). The survivors' fear of a recurrence of a tsunami and the reasons why they did not trust the early warning system are also explored in this chapter, which ends with concluding remarks on resilience building

Finding meaning in ambiguous loss

Resiliency brings more than recovery and provides the crucible in which people become stronger in difficult and uncertain times (Boss 2006: 48–49). People's beliefs are said to be paramount in determining their fragility or resiliency under stress (ibid.: 59). Thai Buddhism's views of life and death played a role in how people dealt with loss of human life after the tsunami (Soorjjakool 2007). From a psychological perspective, individual resilience can be defined as an ability to be mentally flexible and have the psychological elasticity to stretch in response to a disaster without totally losing a sense of equilibrium (ibid.: 48). Resiliency enables a person to function better and possess the capacity to maintain stability after a disaster (ibid.: 55).

Ambiguous loss is difficult to cope with and initially it helps people to deny the risk of permanent loss. However, when hope of finding the missing person alive runs out, there must be a shift in view and an acceptance that the loss is permanent (Boss 2006: 61). Rituals are designed to help people find meaning in loss and thereby restore resilience, but loss rituals are for those who are evidently dead. The counterfeit funeral described in Chapter 5, is an exception and an unusual Buddhist ceremony used in ambiguous loss with the special purpose of communicating with the missing person and locating the body. Finding meaning is difficult when somebody has disappeared or met violent a death, and Boss mentions that seeking justice is commonly considered to be a meaningful action for those close to a missing person (ibid.: 86). In Buddhist thinking, the person's karma justifies the death and I found that monks' explanations about karma and death were experienced as satisfying; thus, families and relatives did not seek further clarifications or justice elsewhere. When the hope of seeing the person again fades away, meaning in life is also lost for those close to the person. That is the most vulnerable state and, here, resiliency stands for the ability to also be prepared for the worst scenario (Boss 2006: 97).

After a disaster, most survivors must face the difficult decision of whether they should stay or migrate to somewhere else. Deep, social, bonding ties that embed people within society motivate them to stay (Chamlee-Wright and Storr 2009; Aldrich 2012: 50). I found that Thai survivors who experienced ambiguous loss were attached to place in a certain way. Similarly to what was found in post-tsunami India, survivors first waited for proof of life or death, and they hoped for a happy reunion. Later, they shifted their focus to hoping for a body to bury and, much later, they hopefully found ways to accept the ambiguity (Boss 2006: 163). In Thailand ambiguous loss acted as a barrier to leaving the area as the survivors felt that they had to stay, otherwise they had abandoned those who were missing.

Relief mission and resilience building by the Asoke group

As soon as people learned of the tsunami catastrophe, both lay and ordained people travelled to the south of Thailand to assist the affected people and the small village temples. Most of the monks who went did so in groups and they stayed at the village temples. But there were also some monks who went to help as individuals and they stayed outdoors in tents or in temporary *kuti*s (huts). One large group that went on a tsunami relief mission was the Buddhist group Asoke, which was mentioned earlier. Despite the fact that the Asoke group is excluded from the Thai *sangha*, it has a strong bonding social capital and a network of centres in all parts of Thailand. Both the ordained Asoke members and the lay Asoke followers pursue an ascetic and environmentally-sustainable lifestyle. The movement has replaced traditional rituals with training of the mind through daily work, frugality and discipline. The Asoke movement rejects materialism and consumerism (Heikkilä-Horn 2003). Its members are socially engaged and, after the tsunami, they organized their own relief mission to tsunami-stricken areas in southern Thailand.

After the tsunami struck, two Asoke members travelled to the south and assessed what the survivors needed.[1] They recruited volunteers, who were carefully selected, and about two weeks after the disaster had happened, more than 850 Asoke monks, ten precept nuns (*sikkhamats*) and lay Asoke followers went to Phang Nga province in buses and cars. They established a base camp in a small town and slept outdoors under umbrellas with mosquito nets, bringing with them what they needed because they did not want to be a burden to the villagers. They worked in small groups of about five or six people, and every person was assigned responsibilities. The Asoke group's relief mission was not about bringing material aid; instead they helped out with practical matters. An Asoke monk explained to me that Buddhism is about social work and Buddhism teaches people how to support themselves. He added that if people lack basic needs, they must help them until they can support themselves and then they teach them how to support others. The Asoke monks talked with the survivors and at the temporary shelters they carried out various tasks such as waste management, maintaining sanitary conditions, cleaning, hygiene control, medical check-ups, haircutting, teaching and so forth. They also helped the survivors repair and clean their houses, and some survivors moved back into their homes after the monks' assistance. An Asoke monk explained that before they embarked on the relief mission they had been trained in what they should focus on, what they could say and what they should not mention that might add to the survivors' existing suffering. The group therefore refrained from asking the survivors about their personal loss. If the survivors told the Asoke members without prompting, then they could talk about victims' personal problems.[2]

The Asoke group kept up their daily monastic routines, including the alms rounds and *dhamma* talks. One lay Asoke member explained that they did all kinds of manual work, including waste management and cleaning the toilets. The survivors appreciated their hard work and expressed confidence in the Asoke group. The survivors came to talk with the Asoke group about their difficulties. Asoke members said that the survivors felt more relaxed and balanced when they had talked with them. The group, in turn. said that they wanted to share the suffering with the survivors so that they could provide them with *metta* (loving kindness). They described their work in the post-tsunami situation as listening to the survivors and helping them to accept what had happened and to go on with their lives.

Most Thai Buddhists are not vegetarians, but the Asoke group emphasizes the importance of eating only vegetarian food. During their relief effort they cooked vegetarian food and invited people to eat, and they also organized sessions for people to learn how to cook vegetarian food. The survivors offered local vegetables that they cooked together. The Asoke members called it "cooking therapy". The survivors concentrated on cooking and could focus on something else besides the tsunami, and the Asoke members said that not talking about the tsunami was positive for the survivors' recovery. Asoke group members stressed the importance of recovering mentally. In their view, there was a lack of balance between material aid and mental recovery.

The relief work after the tsunami was demanding for individual Asoke members. They met in the afternoon every day to share information, to report what they had carried out during the day, to discuss problems and to plan for the next day. The group encouraged the survivors to work together and with their new neighbours, and with volunteers and others who worked in the area. Thereby they increased their bonding and bridging social capital. The Asoke group's support enabled survivors' resilience building. Counselling and *dhamma* talk provided them with tools to handle emotions and suffering and instil some tranquillity. The survivors wanted the Asoke group to stay longer, but the group had decided beforehand that their relief effort should last for about three weeks and then the survivors had to take over the work that they had organized.

Boss compares Buddhist and Western ways of handling suffering and concludes that Buddhism considers suffering as something that inevitably belongs to the human condition. This is helpful for building resilience and relief because the focus is not on the individual and, furthermore, everything is in a process of change. By contrast, Western thinking has a focus on the individual and on independence and self-sufficiency, which emphasizes the individual self. The notion that everything should be like it used to be creates obstacles to resilience and the relief of suffering (Dingfelder 2003: 48; Boss 2006: 96).

Self-help groups

Scholars have shown that, in the long run, disasters do not alter the social fabric of disaster-affected neighbourhoods and, for example, altruistic cooperation is short-lived (Aldrich 2012: 52). I found that the self-help groups that were set up after the tsunami were very important for the members of the groups in building resilience and creating networks and social capital. Resilience is often a function of a person's social support network (Riessman 1997). Many of the women I met in the different self-help groups had lost the social support networks they had before the tsunami and had to create a new sense of belonging. Through these groups, people gained access to resources and created bonding social capital with other group members also links to leaders and NGOs. Observations and interviews with members of the groups have shown that these groups have helped people develop resilience. Survivors explained that the groups became their lifeline for many months after the tsunami. They felt more relaxed when concentrating on the activities in the groups and being together with survivors who had similar experiences. What stands out are the internal factors and how much the survivors have helped each other to regenerate inner strength. Paradoxical as it may seem, giving help has been a way of being helped, and these people developed from being victims to being survivors.

Monks that I have talked with emphasized the importance of helping others in the process of helping oneself. It was Buddhist monks that initiated some of the self-help groups that emerged after the tsunami. More women than men participated in these groups because they focused on handicrafts like sewing and weaving, which attracted women. The Maya Gotami Foundation, led by the Buddhist monk Ajahn Mitsuo Gavesako, brought the Saori weaving project to the tsunami-hit areas.[3] In 2004, he had introduced the Saori weaving technique in Chiang Mai and Kanchanburi provinces as a kind of meditation technique for those who, for various reasons, could not use traditional meditation techniques. I met Ajahn Gavesako when he visited one of the Saori groups in Phang Nga and I asked him about the technique.[4] He stressed that it was a free-weaving technique that anybody could carry out. Ajahn Gavesako said that the weaving helped people to concentrate and it had positive effects on the minds of troubled people. He wanted to help those who were affected by the tsunami, and he decided to see whether this weaving technique could be therapeutic for them. On 7 February 2005, the Saori project was launched in Phang Nga. It has developed from being a mental health project into an enterprise where 62 people earn their living.

Thip is a 35-year-old woman who joined a Saori weaving group with seven looms. She did not want to talk about the tsunami, saying that it made her sad to think about what happened on the day she lost her home, her livelihood and the people who were precious to her. She said that she was happy to be working in the Saori project, and was proud to show what she had woven. She had a small child who was born after the tsunami, who is with her all the time. It has been difficult for her to find a job, and the weaving gives her a regular income. She said:

> I weave most of the day. Sometimes, I talk with friends at the same time. Before I began to weave I became irritated easily, but when I weave I have to be calm. If a thread gets entangled, I have to be calm to get it in order. I have to be relaxed. When I work together with the others in the group, I feel happy. We weave, talk, and support each other.[5]

Ne is 28 years old and is also a member of the Saori project, though not in the same group as Thip. Ne lost both her husband and her child in the tsunami. They were at home and she was at the temple collecting her mother when the tsunami hit. Before the tsunami, she and her husband worked at two of the large tourist hotels. Both hotels were smashed to the ground by the waves. She said that after the tsunami she felt depressed and did not want to meet or talk to anybody. She still took medicine more than two years after the disaster. She said that it was a neighbour at the temporary shelter who brought her to the Saori project. Ne said:

> Before I joined the Saori project I felt totally confused. I did not know what to do with my life; I had lost my husband and daughter in the tsunami. I joined the Saori project because I wanted to work and train to be with other people. Initially, I sewed and now I am weaving. I really enjoy weaving. I like to choose colours and to make patterns. I also like being together with

> the other women. I have changed my plans. My goal is to work at a hotel again but I will continue to work here for a longer period. I appreciate my friends here. I feel more relaxed here than working at a hotel. Another reason is that I have to be close to my mother so I can pick her up if there would be another earthquake. Perhaps I will work at a hotel in the future.[6]

Mina, who we met earlier, joined a sewing group that was set up in a tent close to a temporary "temple" and the temporary shelter where she and her family lived after the tsunami. Mina and the other women in the group had daily contact with that temple. They participated in ceremonies, donated food and went there for advice and counselling. The members in the self-help group considered participation in the ceremonies together with other survivors, contact with the monks and the daily work in the self-help group to be important for recovery. Mina said that the sewing group had been her saviour.[7] A Thai woman who had extensive experience of working at an NGO in Bangkok had initiated and led the self-help group. She arrived in Phang Nga after the disaster and worked on a voluntary basis in the tsunami-hit region. She had strong social capital from her work in Bangkok and she had both national and international networks. At the end of the first year after the tsunami, the self-help group member numbers had diminished and some of them had left the temporary shelter. The sewing group mentioned above survived longer than most of the self-help groups that were set up after the tsunami. The group broke up in October 2007. The members divided the donated sewing machines and other equipment among themselves. Mina and a few of the other members continued to sew at home and that contributed to their livelihood for a couple of years.

A bridge between the NGOs and the survivors

The Buddhist temples in the affected areas received donations from private individuals as well as international and Thai NGOs, so running the projects involved a lot of administrative work. One temple had projects that concerned a wide range of activities, such as house building, boat building, the distribution of fishing equipment, the gathering of funds for education and school books, the establishing of a library for children, and the distribution of milk to infants. The abbot said to me that there are two things that are especially dangerous for Buddhist monks: money and women.[8] He was aware that the activities at the temple would cause both the temple and the monks problems because they involved a lot of money and also laypeople, both men and women. The abbot was particular about keeping records and being transparent about the donations and how they were spent. At one of my first visits to the temple, he showed me the temple projects' bookkeeping records.

This temple is an example of how temples serve as a bridge between the NGOs and the individual survivors. The NGOs provide the resources and the temples distribute them to the survivors. The abbot had a strong social capital through the *sangha* and through his long engagement in social work before the tsunami, and he also had personal contacts with INGOs. The INGOs knew that the monk was experienced in running projects, that he was reliable and that they could avoid corruption by distributing aid under his leadership.[9] These were important issues, since corruption was a widespread problem after the tsunami, and there was no accountability for governmental or other aid providers. Humanitarian organizations depended on the local authorities for information and for access to the affected areas, and they were therefore reluctant to report incidents of corruption by government personnel to their superiors (Human Rights Center and East–West Center 2005: 96). Corruption was so visible to everybody that a relief worker mentioned that a village leader openly admitted taking 20 per cent commission from government tsunami aid that was meant for the villagers, explaining that such funds were needed to lobby officials. A Thai relief worker who was in charge of an international NGOs' relief project had difficulties with the local government officers who requested money for everything that was distributed to the survivors. The foreign NGO did not agree to the officer's request and had to move the relief site, which was costly. The relief worker said:

> I am not alone in considering the local officers and village leaders as a local mafia. Many people think like me. Greedy is a word that I have in mind when I communicate with them. They have tried to get rid of me because I prefer to be honest and I refuse to cheat. I make them angry, but I have to work in the right way. If I had given them what they wanted, our activities would have worked out very well and everything would have been smooth. But I have refused to be part of the corruption, and that has created some problems for the project, but we have dealt with that successfully. I must follow the NGO's policy and we cannot, like some other organizations, just hand over a large amount of money to the village leader and trust him to manage the projects for the villagers. Co-operation with the village leaders is necessary but he and his staff cannot take over the responsibility for the money and the projects.[10]

Most of the projects at the temple mentioned above went well, but not all. One Thai foundation donated money to the temple and assisted more than 30 families from a tsunami-affected island in building new houses on the mainland. A European Christian organization was meant to follow that example by donating houses to more than 40 families who stayed at the temple. However, in the end, the Christian organization revealed that the families had to convert to Christianity in order to receive the donation. This was something that the families, who had lived at the Buddhist temple for more than a year, did not accept. They were Moken and Moklen and had developed close ties with the Buddhist temple. That increased their social capital and they acquired access to resources that they otherwise could not get. However, they were excluded by the villagers who disapproved of them staying at the temple for such a long time.

The temple had been careful about not permitting one family to receive excessive resources, which was said to be the case at other places. In the end, the abbot solved the situation by using his network and social capital and cooperating with some other Christian organizations that donated financial resources so the families could finally move into their new houses.

INGOs and Christian missionaries especially had targeted the Moken and Moklen group and arrived with aid specifically for that group. The Moken and Moklen's excluded and vulnerable situation gave them access to resources but it was at a very high cost. Groups of them converted in exchange for aid, which excluded them from the rest in their ethnic group. Those that converted had to give up performing the important traditional ancestor ceremonies.[11]

Distribution of aid and searching for protection and good fortune

In the long-term aftermath phase, a severe consequence for many of the survivors was that they lost their livelihoods. It was hoped that the lack of jobs and unemployment would be short-term, but for many it was difficult to find an occupation. Moreover, the distribution of compensation did not correspond to the actual damages. Phang Nga's weak social capital compared to Phuket and other provinces' higher level of social capital was made obvious by the government's distribution of aid. Fisheries in Phang Nga that suffered 50 per cent of the total damage received only 30 per cent of the total value of aid. In contrast, Satun's share of damage was only 6.6 per cent but the area received 39 per cent of total compensation. Krabi fisheries, with 65 per cent of total compensation, also received much more than their fair share. One suggested reason for the uneven compensation and low social capital could be that Phang Nga was not given the highest priority because of the many ethnic minority groups in the area. Another reason for unequal disbursement might have been the result of the allocation of budgets by central and local governments (Jayasuriya 2010: 192–193).

The villagers tried various ways to solve their difficult situation. Several fishermen and women suffered from PTSD (Post-Traumatic Stress Disorder) and had hallucinations; they were unable to continue fishing. Many tried to find new occupations but found it difficult. During that time amulets were in great demand and, as a last resort, the villagers believed they would give them luck. Monks that I interviewed who, under normal circumstances, did not hand out amulets said that they used to bring amulets to temporary shelters and distribute them to those villagers who asked for them. The monks explained that the survivors felt protected by the amulets and they gave them comfort and helped to build some resilience. In 2006, the villagers in Phang Nga began wearing the big circular amulet Jatukham–Rammathep, called Jatukham. This amulet was believed to bring the owner wealth and help them in the difficult economic situation after the disaster. For more than a year the amulet became extremely popular, not only in Phang Nga but all over the country (Kitiarsa 2012: 111). The amulet's fame was spread throughout the country by the media and by word of mouth between ordinary people, especially with regard to miracle stories about its efficacy, which made it much sought after. Thus large-scale worship can be related to aspects of psychological, socio-economic and political processes in contemporary Thailand (ibid.: 121).

The amulet originated from Nakhorn Si Thammarat in southern Thailand, a province where many of the villagers who lived in the affected district in Phang Nga came from. It was named after the two princes Jatukham and Rammathep, who are considered to be the founders of Nakhorn Si Thammarat and believed to possess extraordinary supernatural powers. It was also believed that owning a Jatukham–Rammathep amulet would protect the owner as well as provide success in life. The first Jatukham–Rammathep amulet was created in 1987 by a Thai policeman who was convinced that the two princes' spirits had helped him to solve a difficult murder case. After the policeman's death in 2006, the amulet's popularity increased enormously (Kitiarsa 2012). It was distributed by the main temple, Mahathat Woromaha Vihan, in the Nakhorn Si Thammarat province town. Other temples became involved in the production of the Jatukham amulets since the Mahathat Woromaha Vihan alone could not meet the great demand.

How fast the villagers recovered after the tsunami varied from individual to individual. The trauma that was associated with loss of lives, physical assets and livelihood imposed heavy pressures on the survivors. Those who had a slow recovery often had a combination of difficulties such as, for example, suffering from ambiguous loss and finding a sustainable means of livelihood, which was vital for their recovery. The Jatukham amulet was believed to grant good fortune and solve personal problems. At the end of January 2007, I was invited by a family to travel with them to visit the temple Mahathat Woromaha Vihan. It was a five-hour journey by car. It was noon when we arrived in Nakhorn Si Thammarat town and the streets that surrounded the temple were jammed with traffic. The temple area was crowded with people and there were many groups, some of which were led by monks who conducted ceremonies under "nets" of sacred threads (*sai sin*) with loose ends of *sai sin* threads hanging from the "nets" for people to keep in their folded hands during ceremonies. The Jatukham amulets were produced manually outdoors on the temple grounds. People flocked around the "amulet-machines" that imprinted the pattern and text on the blank material. I noticed that some monks were also involved in the amulet production process. We visited several temples in Nakhorn Si Thammarat and at almost every temple lay women sat on the ground, wore masks and sifted the material for the production of the amulets.

The first thing the family did when we arrived at the famous Mahathat Woromaha Vihan temple was to stand in the long queue for paying respect to the temple's relics. After that was done we walked around the temple area and stopped at the amulet booths and observed production of the amulets. The family were convinced about the amulet's power. They related several examples of people that possessed a Jatukham amulet and had better circumstances than they had had before they owned it.

The family I travelled with had struggled greatly after the tsunami. They were traumatized by the loss of family members; they worried constantly about their livelihood and had difficulties making ends meet. Before the tsunami, they had worked in fishing but it was not possible for them to continue fishing after the disaster. They were injured, had lost their boat and their house, and did not dare go back to sea. They had considered moving somewhere else but felt tied to the place since they had lost so many people who were close to them in the tsunami. They said that leaving the place felt like abandoning them. They believed in Buddhism and simultaneously they put their faith in the power of the amulet. The Jatukham amulets were made in different sizes from large to very large. The amulets had different designs, and the price varied. After choosing very carefully, the family bought a set of amulets for themselves to be kept at their small shrine at home. They also bought amulets to wear around their necks and to give to close friends back home.

Worshipping amulets is not in line with the Theravada Buddhist teaching, but faith in objects similar to amulets has a long history in South-East Asia. Most local monks in Phang Nga did not criticize the laypeople's faith in amulets. They assigned the survivors' increasing trust in amulets to their difficult situation and predicted that their faith in these objects would decline when things improved. Phra Payom is a well-known and outspoken monk from central Thailand who visited Takua Pa at the two-year memorial ceremony. He gave a speech at Wat Yan Yao and addressed the faith in the Jatukham amulet in a humoristic manner. He spoke in a sarcastic way about the commodification of Buddhism and stressed the illusion of buying merit and happiness. The lay followers who travelled with the monk had brought several objects for sale. For example, rubber wristbands similar to those people wore for honouring the Thai king were sold here for a much lower price. They also brought cookies with the same shape and design as the Jatukham amulet. Phra Payom said jokingly that Jatukham amulets were expensive and useless but the Jatukham cookies were cheap and nutritious.[12] The villagers laughed and appreciated his talk, but I got the impression that they did not take his speech seriously.

The Jatukham amulet became a lucrative business and it was estimated that in 2007 the amulets generated over 20 billion baht. Eighty million units in 400 batches were produced and circulated in the market (Kitiarsa 2012: 117). Scholars have interpreted the enormous interest in the Jatukham amulet as a sign of societal frustration, a lack of political stability and a desire for spiritual assurance. Kitiarsa understands the Jatukham craze as a growing worldliness of Thai religiosity and an intersection of religious practice with political anxieties (ibid.: 118). The faith in Jatukham could be interpreted as a sign of Buddhism's decline. However, Kitiarsa's viewpoint is that the hybridity between supernatural elements and popular Buddhism could instead be seen as a sign of Buddhism's wellbeing (ibid.). The survivors of the tsunami in Phang Nga who worshipped the Jatukham amulet hoped their difficult situation after the disaster would take a turn for the better. I understand their faith in the amulets as a signifier of a profound crisis after the catastrophe and as a complement to their belief in Buddhism.

The villagers in Phang Nga wore the amulets for less than one year. The family that I accompanied to the temple in Nakhorn Si Thammarat said that they still believed in the power of the amulet but had stopped wearing it and placed the amulets on their shrine at home. In 2008 their situation had improved marginally and had become more stable. An important change was that they had accepted the loss of their family members.

Evoking ancestors' protection

Fulfilling obligations to their ancestors, who in return could provide safety and protection, was a way for people to shield themselves against disasters and misfortunes. At *wan phra* on Wednesday 9 May 2007, a village in Takua Pa had invited a *nora* troupe to conduct a three-day ancestral rite, *nora rong khru.*[13] A *nora* performance is known to many as a folk-dance drama. That is because *nora* performers appear and perform dances and sing songs for entertainment at various gatherings and festivals. In the south of Thailand *nora rong khru* plays a serious role in many people's lives. It is part of a southern ancestral tradition that communicates with ancestors through song, dance and spirit possession (Jungwiwattanaporn 2006: 376). The ancestors give blessings and protection, and have the capacity to heal illness. Most of the *nora* troupes are based in Songkhla, Pattalung, Nakhon Si Thammarat and Trang provinces and they are not so common in Phang Nga.

Many rituals, from daily offerings to large ceremonies such as the *nora rong khru*, are ways to cope with anxieties. The Buddhist method of dealing with worries is to practise mindfulness. The more animistic method is to build friendly relationships between humans, ghosts and spirits (Jungwiwattanaporn 2006: 383). *Nora* followers believe that the ancestral spirits are demi-gods with supernatural powers and they can bless their family members if they are pleased, or punish those that offend them. The *nora* rite makes it possible for the spirits to communicate with their descendants through possessed mediums who are in a trance and the rite gives the living an opportunity to show gratitude to the dead. *Nora rong khru* also serve goals such as vow-fulfilment ceremonies, healing illness and graduation ceremonies for *nora* trainees (ibid.: 382).

The *nora* performance is carried out by visiting *nora* troupes and invitations to such a troupe is costly and has to be planned far in advance. Invitations are commonly made by a family that also becomes the host for the *nora* troupe, but they can also be done collectively by a village through the village temple. In the village in Takua Pa district, laypeople together with the village temple had raised money and invited a *nora* troupe to hold the three-day *nora* ceremony. The area where the temple was located was one of the worst hit by the tsunami and many local people had died. The villagers venerated the late abbot and they told me that the temple was miraculously saved from the tsunami thanks to his power. The tsunami waves had swept past both sides of the temple and left it untouched.

The *nora* ceremony in Takua Pa followed the tradition of having the performance conducted from Wednesday until Friday and it had all the different sequences that a *nora* ceremony contains (Guelden 2005; Jungwiwattanaporn 2006). The main aim of holding the *nora* ceremony in Takua Pa was for people to build good relationships with their ancestors, and this was done by paying respect to the ancestors and entertaining them; in return the villagers would receive the ancestors' protection. Other aims of the *nora* rite were to hold a healing session for sick children and to consecrate a new *nora* medium. Villagers explained that participating in the *nora* ceremony was important for fulfilling vows (*kae bun*) and that the *nora* ceremony served as a protection from future disasters and misfortunes for themselves, their families and the village.[14]

In the south of Thailand, Theravada Buddhism has developed a close relationship to the *nora.* The Buddhist temples and *nora* have such a strong association that some consider *nora* as a Buddhist tradition (Horstmann 2011: 188). The *nora*, however, is open to all faiths and incorporates various elements from other religions (Guelden 2005). It is important to note that one of *nora*'s roles is to connect the ancestors and their descendants with the purpose of settling conflicts among villagers and creating harmony and unity in the villages (Jungwiwattanaport 2006: 375). Different faiths should not be a hindrance to achieving that goal.

In Takua Pa, the day before the *nora* ceremony took place, a group of local women received temporary Buddhist ordination as *chi phram*s. A villager told me that it was common at this temple that women went forth as *chi phram*s the day before *wan phra.* However, on this occasion when the temple also organized the *nora* ceremony, more women than usual had received temporary ordination with the purpose of vow fulfilment.[15]

The Buddhist *tham bun* ceremony is commonly held in the temple's *sala.* However, on the first day of the *nora* ceremony, which was the *wan phra*, the ceremony took place in the special *rong nora* ritual booth. The ritual booth was a simple construction with a grass roof, supported by poles and partly without walls. The monks conducted the *tham bun* ceremony in the same way as usual. In the morning monks, *chi phram*s and laypeople gathered in the ritual booth. The villagers had brought food containers as they used to do on *wan phra*, and they offered food to the monks. Outside the ritual booth was a table with black bowls for money donations. The bowls were the same as those used in other ceremonies at the temple.

The mediator between the mundane world and the spirit world is called *nairong nora.* He arrived in Takua Pa with his troupe of musicians and male and female dancers. The *nora* ceremony began in the afternoon. There was not enough room for all participants in the *nora* ritual booth and therefore people watched the ceremony sitting in a big tent that was placed in conjunction with the booth. Inside the tent there was a shrine with Buddha statues and a photograph of the temple's well-known late abbot, together with photographs of Buddhist ascetics. The position of the late abbot's photograph revealed his elevated status and that made me think of him as the villagers' main ancestor. The villagers paid respect to their ancestors at home by offering food and drink during the three days of the *nora* ceremony.

According to the *nora* performers, the ceremony was crowded with ancestral spirits. A person in trance was possessed by one spirit at a time. Some spirits stayed for a short time while other spirits were more persistent and held onto the mediator for longer. The spirits are known for being unpredictable and could also possess participants in the audience. Before leaving the possessed body, the spirits gave their descendants their blessings. The reunion between the ancestral spirits and the living was positive and the participants tried to figure out who the spirits were that they were lucky to have met during the ceremony. It is common that the spirits act as counsellors for the living and meeting them is very meaningful and therapeutic for the living (Jungwiwattanaporn 2006: 389).

Curing sickness is a common request at *nora rong khru* ceremonies. People come to the *nora* ceremony with all kinds of illnesses. It is believed that a person can become sick because he or she has offended the spirits by neglecting them or, for example, killing a snake that was really an ancestor spirit (Guelden 2005: 128–129). In Takua Pa, the healing rites took place on both the second and third day of the *nora* ceremony. In Takua Pa it was mostly children who were brought to the healing ceremony.

Through the *nora* ceremony it is believed that parents and grandparents who have passed away have not actually gone but are there to protect and support their descendants. That was comforting to the participants and they felt protected and closer to their older relatives. The *nora* ancestors emphasize harmony in the community and warn that the many disputes over land ownership that had occurred after the tsunami would lead to more misfortune. Conflicts had also arisen because of the notion of unfair treatment and assistance after the tsunami, triggering social tension, which the *nora* rites would hopefully reconcile.

Warning and protection

There was no alert system in the affected regions when the tsunami struck. After the catastrophe, the villagers were worried about the threat of a new tsunami. There were numerous rumours of new earthquakes with ensuing tsunamis that increased the stress among the survivors. They had also been evacuated a number of times and most of my informants who lived in reach of a tsunami always had a bag packed with the necessary items they would need in case of an emergency. Almost two years after the catastrophe, a deep-sea buoy was designed to detect tsunami waves and send signals to warning centres in the countries at risk. But the question was, how would they communicate the risk to the people in the coastal areas? The solution was to put up 55 tsunami-warning poles in the affected provinces. The government had put up signs pointing to escape routes and installed loud speakers on the poles. It was debated in the media and among the villagers whether or not the warning system would work and they were not sure if they could trust it.

My assistant and I participated in the preparation for the first evacuation drill and test of the system in Phang Nga. The drill was held on 25 July 2007 and it was the first joint early tsunami warning exercise in all the six provinces that were hit by the tsunami. We had chosen to participate and observe the exercise in Ban Nam Khem and we went there at 8 a.m. One of my informants had participated in the preparation meetings that planned the exercise. The warning drill was to begin at 9a.m. and the organization committee had decided to close some of the roads to traffic to make it safer for the villagers to run. Many officials had gathered at the rescue route. All people that we talked to were very curious about how the early warning system would work. It was rumoured that the copper threads had been stolen from the warning poles and therefore they would not work. By 9.50a.m. our group had waited for the signal for about one hour. We stood very close to one warning pole and we heard a weak signal followed by a message. A voice said that an earthquake had occurred in the ocean and everybody should leave the beach as fast as possible because a tsunami would follow. The message was in Thai, English, German, Japanese and Burmese.

As soon as the people heard the signal they left their houses. Some went on their motorbikes uphill and others walked or ran. Most of the people who participated in the drill were women and children. The school was located on safe ground and the villagers were instructed to go there. At the school, the villagers registered that they had participated in the drill and they received a box with lunch and bottles of water.

In Phang Nga, the warning drill was carefully planned and groups of villagers had had meetings and rehearsed it. Prominent guests were invited: the governor, the district leader and two UNESCO representatives came to the school; one from Paris and one from Bangkok, and the guests held short speeches. The media covered the drill and some villagers were interviewed for the TV channels. One of the American missionaries turned up at the school. He talked to the guests from UNESCO Paris and UNESCO Bangkok and told them that the warning signals did not work. The missionary lived in a house close to the sea. He and his family had closed the door to their house and they could not hear the signal at all. The missionary drew the conclusion that if an earthquake were to happen in the middle of the night and if it rained, they would not be woken up by the signal. It was a clear day with almost no wind when the evacuation drill was conducted but the family could not hear the signal although they were waiting for it and tried hard to hear it.

The Thai villagers were not confident about the early warning system either and the insecurity that the survivors felt about it, together with the concerns they had about a possible new disaster led them to find ways to protect themselves. Initially, most of my informants did not dare to sleep in their houses unless somebody watched over the sea. The villagers organized themselves in groups and took turns to keep guard. They called each other on their mobile phones if they observed something that could be a sign of a tsunami. The early warning system has been improved and the signals have become louder; over time, the villagers have gained more trust in it than they had in the beginning.

Concluding remarks

This chapter has dealt with local initiatives and factors that were important in processes of resilience building and post-disaster recovery. Disasters devastate fundamental conditions for people's lives including routines, livelihoods and valued belongings. Human resources and facilities in the immediate local community are essential in the acute phase after the disaster. Social networks provide important information, support and guidance. People who are well connected and have high levels of social capital can, through their networks, act and raise support while others wait for guidance from officials. Aldrich found that post-disaster social networks are likely to closely mirror pre-disaster conditions, and changes to pre-disaster social capital were limited in time and effect (Aldrich 2012: 53).

The *sangha*'s high level of social capital was evident in post-tsunami Thailand and it was used to assist the survivors both with their mental survival and with their material needs. The strong bonding social capital among the villagers created negative outcomes for the Moken and Moklen group, who were not welcome in the village and were considered to have overstayed the temple's hospitality. The Buddhist group, mentioned in this chapter, that is excluded from the *sangha* managed to support the survivors in their recovery and worked efficiently thanks to their network and skilled management. The *sikkhamat*s are included as an integral part of that Buddhist group and they worked side by side with the monks and the lay followers. Their expressed goal was to help the survivors mentally, to give them hope and build resilience.[16]

After a disaster, religious beliefs commonly give the survivors a sense of control. Rituals and ceremonies have proved to have healing capacities both individually and collectively (Ellison and Katz 2010: 188–192). Johnson *et al.* have found that rituals create a safe and contained space in which to face emotions where trauma is recognized, and that they could become metaphors for transformation and change. Rituals and ceremonies embody attachment to the family, the community and society, and reflect that we are united in the journey away from the disaster (Johnson *et al.* 1995). Furthermore, rituals reinforce bonds between individuals and assist adaptation by creating security and warding off threats (Danieli and Nader 2006). The increased participation in Buddhist ceremonies after the tsunami, the faith in the Jatukham–Rammathep amulet and the *nora rong khru* ancestral rites are, as I see it, people's ways of coping and regaining control of their lives and over an unsure destiny. The findings presented here support the notion that religious practices, social capital, networks and self-help groups are building blocks of resilience. Both men and women were active in finding ways to support themselves. More women than men participated in the self-help groups that were organized and those groups provided social networks, increased social capital and built the members' resilience. In the long-term aftermath phase, the majority of disaster-affected people usually cope sufficiently, but there are also those who remain distressed and functionally impaired (International Work Group on Death, Dying and Bereavement 2002: 460). Loss of livelihood was, among my informants, not experienced as a main problem directly after the tsunami. It was later in the aftermath phase that most survivors first grasped how extensive and long-lasting the loss of livelihood was. Many survivors became uncertain about their future and the loss of a source of income was experienced as beyond their control. Mental health research showed that the loss of livelihood as a result of the tsunami was independently and significantly associated with all mental health conditions (Van Griensven *et al.* 2006: 547).

Notes

1 Interviews with Asoke monks, *sikkhamat*s and Asoke members were conducted in October 2005 and August 2007.
2 Ibid. and interviews with lay Asoke followers in August 2007.
3 The Maya Gotami Foundation was founded by Ajahn Gavesako in 1990 and runs a variety of educational and other social projects.
4 Interview, 3 February 2007.
5 Interview, 30 March 2007.
6 Interview, 27 March 2007.
7 Interview, 27 March 2007.
8 Interview, April 2006.
9 Informal talk with INGO workers.
10 Interview, 8 September 2006.
11 Information from visiting the islands
12 The speech was held on 26 December 2006.
13 *Nora* shortened from Manohra.
14 Personal communication, 9 May 2007.
15 Personal communication, 10 May 2007.
16 Interview with a *sikkhamat*, October 2005.

References

Aldrich, Daniel P. (2012) *Building Resilience: Social Capital in Post-Disaster Recovery.* Chicago and London: The University of Chicago Press.

Boss, Pauline (2006) *Loss, Trauma, and Resilience: Therapeutic Work with Ambiguous Loss.* New York and London: W.W. Norton & Company.

Brenner, H. Grant, Daniel H. Bush and Joshua Moses (eds) (2010) *Creating Spiritual and Psychological Resilience: Integrating Care in Disaster Relief Work.* New York and London: Routledge.

Chamlee-Wright, E. and V. Storr (2009) "Club Goods and Post-Disaster Community Return". *Rationality and Society* 21 (4): 429–458.

Danieli, Y. and K. Nader (2006) "Respecting Cultural, Religious and Ethnic Differences in the Prevention and Treatment of Psychological Sequelae". In L. Schein, H. Spitz, G. Gurligame and P. Muskin (eds) *Group Approaches to Terrorist Disasters.* New York: Haworth Press, pp. 203–234.

Dingfelder, S. (2003) "Tibetan Buddhism and Research Psychology: A Match Made in Nirvana?" *Monitor on Psychology*, 34(11): 46–48.

Ellison, Koshin Paley and Craig L. Katz (2010) "Rituals, Routines, and Resilience". In *Creating Spiritual and Psychological Resilience: Integrating Care in Disaster Relief Work*, H. Grant Brenner, Daniel H. Bush and Joshua Moses (eds) New York and London: Routledge, pp. 181–193.

Guelden, Marlene P. (2005) *Ancestral Spirit Mediumship in Southern Thailand: The* NORA *Performance as a Symbol of the South on the Periphery of a Buddhist Nation-State.* Unpublished dissertation, University of Hawai'i.

Heikkilä-Horn, Marja-Leena (2003) "Small is beautiful in Asoke villages". In *Insights Into Santi Asoke*, Marja-Leena Heikkilä-Horn and Rassamee Krisanamis (eds) Bangkok: Fah-apai Co Ltd, pp. 25–62.

Hill, P. C. and Kenneth I. Pargament (2003) "Advances in the Conceptualization and Measurement of Religion and Spirituality: Implications for Physical and Mental Health Research". *American Psychologist*, 58(1): 64–74.

Horstmann, Alexander (2011) "Reconfiguring Manora Rongkru: Ancestor Worship and Spirit Possession in Southern Thailand". In Endres W. Kirsten and Andrea Lauser (eds) *Engaging the Spirit World: Popular Beliefs and Practices in Modern Southeast Asia*, New York and Oxford: Berghahn Books.

Human Rights Center, University of California Berkeley and East–West Center (2005) *After the Tsunami: Human Rights of Vulnerable Populations*. Berkeley, CA: University of California Press.

International Work Group on Death, Dying and Bereavement (2002) "Assumptions and Principles about Psychosocial Aspects of Disasters". *Death Studies*, 26: 449–462.

Jayasuriya, Sisira and Peter McCawley (eds) (2010) *The Asian Tsunami: Aid and Reconstruction after a Disaster*. Caltenham, UK and Northampton, MA: The Asian Development Bank Institute and Edward Elgar Publishing.

Johnson, David Read, Susan C. Feldman, Hadar Lubin and Steven M. Southwick (1995) "The Therapeutic Use of Ritual and Ceremony in the Treatment of Post-Traumatic Stress Disorder". *Journal of Traumatic Stress*, 8(2): 283–298.

Jungwiwattanaporn, Parichat (2006) "In Contact with the Dead: *Nora Rong Khru Chao Ban* Ritual of Thailand". *Asian Theatre Journal* 23(2): 374–395.

Kitiarsa, Pattana (2012) *Mediums, Monks, & Amulets: Thai Popular Buddhism Today*. Chiang Mai: Silkworm Books.

McCreight, Robert (2010) "Resilience as a Goal and Standard in Emergency Management". *Journal of Homeland Security and Emergency Management* 7(1): 1–7.

Norris, Fran H., Susan P. Stevens, Betty Pfefferbaum, Karen F. Wyche and Rose L. Pfefferbaum (2008) "Community Resilience as a Metaphor, Theory, Set of Capacities, and Strategy for Disaster Readiness". *American Journal of Community Psychology*, 41(1–2): 127–150.

Olshansky, R. (2006) "San Francisco, Kobe, New Orleans: Lessons for Rebuilding". *Social Policy*, 36(2): 17–19.

Pargament, Kenneth I., Harold G. Koenig, and Lisa M. Perez (2000) "The Many Methods Of Religious Coping: Development and Initial Validation of the RCOPE. *Journal of Clinical Psychology*, 56: 519–543.

Riessman, Frank (1997) "Ten Self-Help Principles". *Social Policy*, 27, pp. 6–11.

Sorajjakool, Siroj (2007) "Tsunami and Ghost Stories in Thailand: Exploring the Psychology of Ghosts and Religious Rituals within the Context of Thai Buddhism". *The Journal of Pastoral Care & Counseling*. Winter 61(4): 343–349.

Van Griensven, Fritz, M.L. Chakkraband, W. Thienkrua, W. Pengjuntr, Cardozo B. Lopes, P. Tantipiwatanaskul, S. Ekassawin, A. Varangrat, C. Gotway, M. Sabin and J.W. Tappero (Thailand Post-Tsunami Mental Health Study Group) (2006) "Mental Health Problems Among Adults in Tsunami-Affected Areas in Southern Thailand". *JAMMA*, 2 August, 296(5): 537–548.

10 Conclusion

This monograph has dealt with the aftermath of the 2004 Indian Ocean tsunami with special focus on culture and religion in the recovery process. It has covered responses from affected local communities mainly in Phang Nga province in Thailand, with emphasis on religion's positive as well as negative sides. The aim has been to provide a deeper understanding of how local women and men experience and deal with disasters and how culture and religion interplay with resilience and gendered vulnerability. Buddhist and spiritual beliefs have been found to aid recovery and the study gives ample examples of how Thai Buddhist monks, nuns and laypeople employ Buddhism in order to cope after losing family members, friends, neighbours, their homes and their means of livelihood.

Studies following disasters that have examined recovery and coping have shown that religion is frequently used to assist coping and often brings comfort and hope, which enables survivors to heal and move on with their lives (Pargament 1997; Ellison and Katz 2010; Koenig 2006). Groups that are more religious may derive greater benefits from religion, as well. Religious and spiritual beliefs may represent a profound source of comfort, meaning and hope following disasters for many people, but not for all. Those who have experienced a disaster can also be spiritually injured, especially if they have had faith in being protected by religion and now feel abandoned. They lose certainty in their beliefs and their search for meaning has not provided any answers (Koenig 2006: 10–11). Severe traumatic stress can shatter victims' basic assumptions about themselves and their world, including what they believe in. Christians might question God's love for them and why God let the disaster happen; they might also lose faith in God's power (Pargament *et al.* 2001). On the other hand, intact religious beliefs can help facilitate recovery from severe trauma. Stress-related growth may occur when an individual has experienced a stressful trauma and successfully dealt with the situation. Such persons may be more resilient when facing future difficulties (Reich *et al.* 2010: 312). Religious involvement in the recovery process has been shown to predict greater stress-related growth among people who have experienced extreme stress (Koenig 2006: 11).

Buddhists believe that the suffering they are experiencing is inevitable and created by themselves in this or in previous existences. Since there is no creator and no God in Buddhism, there is nobody that can be blamed for the catastrophe that devastated so many people's lives. A mental health study that was carried out about two months after the tsunami struck showed that the prevalence of PTSD symptoms in Phang Nga after the tsunami was lower than in other populations affected by natural disasters (Van Griensven *et al.* 2006: 547). The differences were ascribed to variances in duration and severity of disasters, and were thought to be specific to coping mechanisms based on cultural differences. Van Griensven *et al.* stated that their findings are consistent with results from other studies showing that Buddhism had a protective effect on anxiety and depression. The Buddhist notion that life is, to a certain extent, predetermined and a result of one's own actions, helped people to accept and overcome negative events in their own lives (ibid.).

Despite the fact that every disaster is unique and there is a great variation in individuals' ability to cope, the aftermath of disasters appears to follow a certain pattern that can be divided into phases (International Work Group on Death, Dying and Bereavement 2002: 451–452). The early aftermath phase is characterized by chaos, especially when communities and individuals lack preparedness for disaster. Psychological needs are often neglected during this phase, and research has found that, besides the urgent physical assistance, the best way to help is to explain and give information, reassurance and instructions (Koenig 2006: 1–3). Thailand was unprepared for a disaster when the tsunami struck and the coordination between the emergency relief organizations was defective.

This study has found that the roles played by the Buddhist temples and institutions in the wake of the disaster were invaluable, and individual Buddhist monks and nuns carried out extremely difficult tasks. In small communities like those in Phang Nga province, the local temples are places for gatherings and religion is intertwined in daily life. The temples were a place where many of those who escaped from the tsunami sought sanctuary. They became crowded with people who stayed in the temple buildings and when these were filled to capacity, the survivors took refuge in tents in the temple grounds. Later, many temporary shelters were set up in connection with the temples. The monks provided leadership, counselling and distributed donated food and material necessities to the survivors. The temples were open to all, irrespective of faith, ethnicity, gender and position in society, and were not only a place to stay for those who had survived the tsunami; many dead bodies were brought there and some temples became major forensic sites. Moreover, temples accommodated many of the Thai volunteers who arrived to help after the disaster. The temples became also the obvious place for news about missing people, communication and information in general.

A disaster is a thoroughly traumatic experience; the reactions of people and their ability to cope are, among other things, dependent on its severity, society's and individual's preparedness for disasters and an individual's personality. After the tsunami, the Buddhist monks became vital in the handling of death and suffering. In Thailand, as in most Asian countries, it is common to categorize a death as either "good" or "bad" and a bad death must be handled with great care that needs the special expertise that the Buddhist monks provide (Tambiah

1970). A violent and/or an untimely death is a bad death and is considered to be a disruption and a threat to the social order, which has to be dealt with carefully. This study shows that the belief in ghosts and spirits is interwoven into the social fabric of the living and, for reasons of safety, it needs to be attended to in ceremonies that are conducted by the monks. Several chapters in this book have shown how vital it was for the relatives to be able to communicate across boundaries between the living and the dead. The communication was carried out in rituals and in dreams and the Buddhist monks were the vehicle for the communication. For many survivors it was a comforting and necessary part of the recovery process to be in contact with deceased family members and to look after their needs. This study found that the Buddhist monks' explanations about life, death and suffering helped the survivors to understand and create meaning from what had happened. The monks and nuns encouraged them to accept what had occurred, not be stuck in the past, and to go on with their lives and help others who had survived.

In the phase when the rescue workers begin to withdraw from the disaster area and the survivors come to realize the broader implications of the disaster and their need for support, grief over their losses becomes central. In this phase there is often a need to attribute meaning to why the disaster happened and why some people survived while others died (Koenig 2006: 3). Religious and spiritual explanations are often sought after and religion is commonly valuable, since it also deals with processes of individuals' integration of the catastrophe and acceptance. In Thailand the long-term aftermath phase was characterized by the fact that many survivors had integrated the trauma into their lives. This study shows that rituals, memorials and commemorative events to honour those who perished in the disaster were extremely important but became detrimental when they were handled without considering local people's needs. Most people adapted to the new situation and some continued to experience emotional distress and remained impaired, especially those who experienced ambiguous loss.

It was common that survivors worried and felt anxious about basic needs in their life, such as their livelihoods. Many survivors who had experienced great loss had difficulties in reorienting themselves and finding new occupations. In the long-term aftermath phase, assistance to restore their livelihoods would have made a great difference and improved the recovery and wellbeing of many the survivors who searched for new ways to make a living.

Gender, vulnerability and resilience building

There are links between social networks, social capital, vulnerability reduction and resilience building (Aldrich 2012). Social vulnerability has become a policy-relevant area and Aguirre argues that vulnerability points to the need for systems to change. It is therefore imperative to identify the key networks to be included in the resilience and mitigation project and mobilize and coordinate them to bring about the required changes (Aguirre 2006: 4–6). The tsunami-affected villages in Thailand have been described as both vulnerable and resilient. Adaptive capacities are related to the acquisition of networked resources and social capital, which is important for reducing vulnerability. After the disaster, the survivors used their local networks and assisted each other. Most survivors were in shock after the catastrophe and initially their main focus was to search for missing family members, relatives and friends. The strength of local agency was apparent, but the magnitude of the tsunami was enormous and the survivors needed help from the government and relief organizations.

Oliver-Smith has stated that disasters are totalizing events, as they unfold all dimensions of a social structural formation, and the totality of its relations with its environment may become involved, affected and focused (Oliver-Smith 1999: 20). Phang Nga had, when the tsunami hit, quite recently developed into a growing tourist destination and the livelihoods of people in the villages close to the Andaman Sea had become increasingly tourism-based (Calgaro *et al.* 2009). It transpired that this development left the local people vulnerable. The expanding tourist industry attracted migrant workers on a large scale from other parts of Thailand and also labourers from neighbouring countries, particularly Myanmar. In Phang Nga, about 80 per cent of the large numbers of Myanmar migrant workers were unregistered and they were among the most vulnerable groups (Hulme 2005). Many of the survivors were dependent on the tourist industry and a report states that tsunami vulnerability in the worst hit tourist destinations is a product of multiple dynamic and interacting factors, including geographical exposure, destination-specific development characteristics, social structures and governance (Calgaro *et al.* 2009: ix). It is noted that the tsunami-hit regions were vulnerable and the small fishermens' livelihoods and political voices in southern Thailand had been eroded (Lebel *et al.* 2006: 132). Sustainable livelihoods reinforce a community's inherent strengths when facing difficulties such as natural hazards. Social cohesion, mutual support, strong societal networks and social capital are vital for human resilience and are often dependent on gender, religion, caste or ethnicity. Spontaneous self-help groups regularly spring up in the immediate aftermath of a disaster. In Thailand, more women than men participated in the self-help groups. Women's agency has proved to be of great importance in relief work and women are described as demanding more support from governments; this is an example of a shift from a vulnerability-based approach towards an agency-based approach, where local communities act and build resilience and achieve sustainability for the future (Smith and Petley 2009).

Among the villagers in Phang Nga, resilience was developed through social networks and webs of relations between people, self-help groups and the Buddhist temples. Bonding, linking and bridging social capital were used and networks operated within and across institutions including the family, religion, politics, economy, medicine and health, education, law and risk management. Social networks can be coordinated, supported and strengthened by social capital at all levels to build resilience, reduce social vulnerability and transform society (Aldrich 2012). After the tsunami, self-help groups emerged and most of them were set up by individual initiatives or by NGOs. Buddhist monks started some self-help groups and more women than men participated in them. From what I have seen, self-help groups that operate over a limited period of time have the capacity to develop resilience in women and belonging to these groups is of vital importance for them. Through bonding, social capital and links to families and relatives, impoverished women also had great coping mechanisms. Such strong social networks helped them survive the negative effects of disruptive shocks ranging from personal threat to large-scale catastrophe. Resilience and resilience-building strategies depend upon capacities and the development of individual or network resources. A person's existing social support network is often seen as vital for resilience (Riessman 1997). However, many of the women I met had lost the social support networks they had before the tsunami. Most of them managed to create new belongings and many did this through the self-help groups.

The differential vulnerability of women and men to disasters is well established (Enarson and Chakrabarti 2009; Phillips and Morrow 2008). Women's vulnerability, compounded by their social inequality and lack of access and control over economic and political resources, is understood. For example, most of those killed by the Indian Ocean tsunami across India, Indonesia and Sri Lanka were women: their increased vulnerability was due to factors such as livelihood roles, being on the beaches while the men were away at sea when the waves hit the coast, and not having physical survival skills such as swimming or tree climbing (Oxfam 2005). The focus of most post-disaster damage and vulnerability assessments, therefore, is rightfully concentrated on the issues that affect women (ibid.: 30).

Gender inequality and vulnerability are closely intertwined (Enarson 2012). Women were particularly at risk of exploitation, violence and discrimination in the tsunami-affected areas. Female survivors who were the most vulnerable included women who were destitute, widows, internally displaced, disabled, Thai and non-Thai migrant workers and those severely traumatized by the disaster and loss of loved ones. Many women who lost their husbands expressed concerns about their future livelihoods. Many women considered the programmes that were set up by authorities and NGOs as something done only to keep women occupied rather than provide jobs.

Thai society is generally noted for relatively egalitarian gender relations arising from women's roles in finance, land ownership, kinship structures and their position in the labour market. Similarly, practices such as equal inheritance of land by sons and daughters, marriage payments from the groom to the bride's family and a preference for post-marital residence with the wife's parents, provide many women with access to economic resources and the emotional support of close family throughout their lives. Yet Thailand is in many ways a highly androcentric society, not least ideologically. Most striking is the fact that conflicting representations of strong, competent women exist alongside the obvious evidence of inequalities and discrimination against them. In post-tsunami Thailand, the government treated men and women differently. The government awarded higher compensation for a dead man than for a dead woman, referring to men as the breadwinners. In reality, Thai women are equally important as breadwinners in society and often bear the entire economic responsibility for children. However, it is not unusual that men and women share the burden of many domestic tasks and both men and women work outside the home. While men, as husbands and fathers, are the recognized "heads" of households and occupy most positions of community leadership, women nonetheless exercise considerable economic and personal autonomy in their roles as respected mothers and the managers of household finances. Women who lost their husbands were discriminated against by receiving no compensation for their missing fishing boats. The compensation went only to men, despite the fact that it is not uncommon for husbands and wives to fish together. The authorities did not acknowledge women as fishers and saw, therefore, no need to compensate widows for the boats they lost in the tsunami. Of those that I interviewed, nobody had insured their boats or other property. Women thereby lost an opportunity to earn their livelihood.

Religion as a resilience factor

Religion is particularly important in times of crises; it is commonly linked to the search for meaning behind disasters and it promotes a social interconnectedness, which is manifested in different ways through the Buddhist *sangha.* Buddhist temples, monks and nuns were crucial as sources of refuge for the survivors in the aftermath of the catastrophe. Religious explanations and the daily and special religious practices were of major significance after the disaster and ceremonies formed an essential part of the survivors' recovery process. Wilson (1989) has shown that rituals, religious ceremonies and other such events often help traumatized people in the healing process. Successful coping strategies reduce the probability that people will undergo mental health problems arising from the experience. Religions have practices and rituals that help adherents to cope, some of which are enacted in groups or individually.

After the tsunami, the *sangha* mobilized quickly and sent forth monks from the Buddhist universities in Bangkok and from many temples. This was done through the *sangha*'s nationwide network of temples and monks, who organized themselves into teams and went to the affected provinces. They worked in groups and supported the survivors and gave assistance at the local temples. The *sangha* proved to be an efficient religious institution. The Buddhist monks' response to the disaster demonstrates the *sangha*'s resilience through their high level of social capital. However, a weak point in the Thai *sangha*'s performance is the gender inequality and the *sangha*'s persistent determination to not recognize any female member of the *sangha.* The Thai Buddhist *sangha* could be even more resilient and more capable in assisting the lay community if female ascetics were acknowledged and fully accepted in their religious role.

Religion had both positive and negative impacts on the survivors' lives after the tsunami. Buddhism became a sanctuary for many survivors and a vital factor in the healing process. However, religion also proved to be destructive and an obstacle in the recovery phase, especially in relation to the Christian missionaries' conversion activities in the area. After the tsunami, the influx of Christian missionaries to the south of Thailand was conspicuous and some missionaries took advantage of the survivors' vulnerable situation, in which they had lost not just family members, relatives and friends, but also their homes and their material necessities. The missionaries arrived well equipped with supplies and arguments and traded material aid for conversion to Christianity. That caused anxiety for many of the converts and meant that families and relatives could no longer be together in the same way as before. Some converts were excluded, due to their new religious belonging, from the community-based and family-centred Buddhist ceremonies and activities that might have provided them with culturally accepted coping mechanisms.

Focusing on the practice of social relations displays how systems of social support and social capital are built and how social relations form the foundation of social networks and shared social resources. This in turn gives an understanding of how resilience building occurs. This study has shown that Buddhist monks, nuns and the temples have been important for networks and for building resilience. Ritual and religious practice play significant roles in resilience building, both in terms of individual and communal capacities to prepare for, withstand and adapt to changed conditions after a disaster. Religious practices develop a sense of connection and wellbeing, and after a disaster it is of special importance to maintain the feeling of belonging and to preserve social relationships. More women than men participated in the Buddhist ceremonies, however women's contributions were not seen as individual, since they participated in the ceremonies in the name of the whole family. Inter-ethnic cooperation was present in some rituals after the tsunami and they had already developed into a hybrid form before the catastrophe. A ceremony like the "counterfeit funeral", mentioned earlier, was seldom heard of prior to the tsunami but became important for those who could not find their family members and relatives after the tsunami and believed that they were dead. This ceremony helped the survivors to deal with their suffering and begin their mourning, which is significant for recovery.

The findings presented in this study support the notion that religious practices, networks and self-help groups play a part in resilience building. I argue that women's active participation in religious activities decreases vulnerability and contributes to resilience building. Religion is often overlooked in research on disasters and I have found that the *sangha* and the Buddhist temples have provided key networks that are vital for building social capital and resilience and for mobilizing mitigation projects. Religious and spiritual beliefs have proved to be vital elements of resilience, and religion is considered to have a positive impact on coping with stressful situations as well as answering meaning-of-life questions (Glicken 2006: 23; Oliver-Smith 1996; and Pargament 2010: 195). Oliver-Smith points out that it is often the case in extreme conditions, particularly those characterized by loss and change, that human beings find themselves confronted with difficult existential questions. The responses of disaster-stricken people invariably involve the moral and ethical core of the belief system and include a deep delving into concepts of both social and cosmic justice (Oliver-Smith 1996: 308). Pargament has found that religion is generally a resilience factor for people who have experienced traumatic events, but he also adds that religion can obstruct recovery. He refers to people who may become angry with God or feel that God has abandoned them or is punishing them (Pargament 2010: 205). Pargament refers to research that has used the Negative Religious Coping Scale, showing that negative religious coping results in a lower overall, existential and psychological quality of life (ibid.: 204). Buddhists do not believe in God and the Buddhists in Phang Nga did not blame anybody for their suffering and misfortune; instead they believe that a person's lifespan and fortune depends on karma. Natural hazards were understood by my informants as something caused by nature. They saw karma as fuel that they had created themselves in this or earlier existences. It was karma that decided when their life was over. Families, relatives, monks and nuns became significant in helping people to form as good circumstances as possible for the next existence.

References

Aguirre, Benigno, E. (2006) *On the Concept of Resilience.* University of Delaware Disaster Research Centre.

Aldrich, Daniel P. (2012) *Building Resilience: Social Capital in Post-Disaster Recovery.* Chicago and London: The University of Chicago Press.

Calgaro, Emma, Kannapa Pongponrat and Sopon Naruchaikusol (2009) *Destination Vulnerability Assessment for Khao Lak, Thailand: Sustainable Recovery and Resilience Building in the Tsunami Affected Region.* Stockholm: Stockholm Environment Institute.

Ellison, Koshin Paley and Craig L. Katz (2010) "Rituals, Routines, and Resilience". In *Creating Spiritual and Psychological Resilience: Integrating Care in Disaster Relief Work*, H. Grant Brenner, Daniel H. Bush and Joshua Moses (eds) New York and London: Routledge, pp. 181–193.

Enarson, Elaine (2012) *Women Confronting Natural Disaster: From Vulnerability to Resilience.* Boulder, London: Lynne Rienner Publishers.

Enarson, Elaine and P.G. Dhar Chakrabarti (2009) *Women, Gender and Disaster: Global Issues and Initiatives.* New Delhi, California, London and Singapore: Sage Publications.

Glicken, Morley D. (2006) *Learning from Resilient People: Lessons We Can Apply to Counselling and Psychotherapy*, London and New Delhi: Sage Publications, pp. 23–26.

Hulme, John (2005) "Forgotten Tsunami Victims: Burmese Immigrants in Thailand". World Socialist Web Site: News & Analysis: Asia. Available at: www1.wsws.org/articles/2005/feb2005/burm-f02.shtml (accessed 4 June 2014).

International Work Group on Death, Dying and Bereavement (2002) "Assumptions and Principles about Psychosocial Aspects of Disasters". *Death Studies*, 26: 449–462.

Koenig, Harold, G. (2006) *In the Wake of Disaster: Religious Responses to Terrorism and Catastrophe.* Philadelphia and London: Templeton Foundation Press.

Lebel, L., S. Khrutmuang and J. Manuta (2006) "Tales from the Margins: Small Fishers in Post tsunami Thailand". *Disaster Prevention and Management*, 15(1): 124–134.

Oliver-Smith, Anthony (1996) "Anthropological Research on Hazards and Disasters". *Annual Review of Anthropology*, 25: 303–328.

Oliver-Smith, Anthony (1999) "'What is a Disaster?': Anthropological Perspectives on a Persistent Question". In Anthony Oliver-Smith and Susanna Hoffman (eds) *The Angry Earth: Disaster in Anthropological Perspective.* New York and London: Routledge, pp. 18–34.

Oxfam (2005) The tsunami's impact on women. *OXFAM Briefing Note* (March).

Pargament, Kenneth, I. (1997) *The Psychology of Religion and Coping: Theory, Research, Practice.* New York and London: The Guilford Press.

Pargament, Kenneth I., Harold, G. Koenig, N. Tarakeshwar and J. Hahn (2001) "Religious Struggle as a Predictor of Mortality among Medically Ill Elderly Patients: A Two-year Longitudinal Study". *Archives of Internal Medicine* 161: 1881–1985.

Pargament, Kenneth I. and Jeremy Cummings (2010) "Anchored by Faith: Religion as a Resilience Factor". In John W. Reich, Alex J. Zautra and John Stuart Hall (eds) *Handbook of Adult Resilience*, New York: Guilford Press, pp. 193–210.

Phillips, Brenda D. and Betty Hearn Morrow (2008) *Women and Disasters: From Theory to Practice.* USA: Xlibris.

Reich, John W., Alex J. Zautra and John Stuart Hall (eds) (2010) *Handbook of Adult Resilience.* New York and London: The Guilford Press.

Riessman, Frank (1997) "Ten Self-Help Principles". *Social Policy*, 27: 6–11.

Smith, Keith and David N. Petley (2009) *Environmental Hazards: Assessing Risk and Reducing Disaster.* New York: Routledge.

Tambiah, Stanley, J. (1970) *Buddhism and the Spirit Cults in North-East Thailand.* Cambridge: Cambridge University Press.

Van Griensven, Fritz, M.L. Chakkraband, W.Thienkrua, W. Pengjuntr, B. Lopes Cardozo, P. Tantipiwatanaskul, S. Ekassawin, A. Varangrat, Sabin C. Gotway and J.W. Tappero (Thailand Post-Tsunami Mental Health Study Group) (2006) "Mental Health Problems Among Adults in Tsunami-Affected Areas in Southern Thailand". *JAMMA*, 2 August, 296(5): 537–548.

Wilson, John P. (1989) *Trauma, Transformation and Healing: An Integrative Approach to Theory, Research and Post-Traumatic Therapy.* New York: Brunner/Mazel.

Selected glossary of Thai, Pali and Sanskrit terms

A note on language and names

Thai is a tonal language, which has a number of vowel and consonant sounds that are not found, for instance, in English. While Thai has its own phonetic script (derived from ancient Indian scripts via mediaeval Khmer), there is no generally established way of romanizing Thai words. In this book, words and phrases in italic typeface are transliterations from standard Thai and rendered phonetically. The transliteration system is a modified version of the Haas convention from 1964 for transcribing Thai. To those who can speak Thai, this will still give sufficient information to indicate which words are meant. For reasons of simplicity, tonal markings and vowel length are excluded, and diacritics have been kept to an absolute minimum. Names and titles are romanized according to established convention, and personal names follow the preferences of the persons concerned. The Thai and Pali words are italicized and may take forms for plurals, e.g. *mae chi*s or adjectives, e.g. *kamm*ic. The orthography of the Pali and Sanskrit words follows the conventions of romanization but omits all diacritical signs.

Abhitham (Thai) ***Abhidhamma*** (Pali) Literally the "higher teaching": the predominantly philosophical and analytical final section of the Tipitaka.

Atthithan (Thai) Making a vow.

Baht (Thai) Thai unit of currency.

Bap (Thai) Wrong, sinful, demerit.

Bangsukul (Thai) ***Pamsukula*** (Pali) Ceremonial rite performed by the monks in honour of the dead.

Barami (Thai) Merit and "power". Perfection, qualities leading to Buddhahood.

Bhikkhu (Pali) Male Buddhist monk.

Bhikkhuni (Pali) Female Buddhist monk.

Binthabat (Thai) Daily alms round.

Bot (Thai) The consecrated building in a monastery where important religious ceremonies and rituals are performed.

Bun (Thai) Buddhist merit.

Chi phram (Thai) A woman with "lay", temporary ordination (without shaving her head).

Dhamma (Pali) Word with many meanings; here it generally refers to the Buddhist view of the nature of things, or the teachings of the Buddha.

Dukkha (Pali) Dissatisfaction, which is one of the three characteristics of existence.

Dana (Pali) Almsgiving, offering, charity, generosity, benevolence, donation, benefaction.

Kamma (Pali), ***karma*** (Sanskrit), ***kam*** (Thai) Intentional or volitional actions, which bring about moral consequences for the performer.

Kilesa (Pali), ***kilet*** (Thai) Moral defilement or impurity, which leads to suffering.

Kuti (Thai) Monks' or nuns' living quarters; often a small basic one-room house.

Lokiya (Pali) Mundane, worldly; associated with the realm of attachment, craving and suffering.

Lokuttara (Pali) Supramundane, transcendental; associated with *nibbana* or the path leading to the attainment of *nibbana.*

Luk chang. "Labourers".

Mae chi (Thai) Term used for white-robed Buddhist female ascetics in Thailand.

Mara (Pali) Buddhist personification of evil.

Metta (Pali) Loving kindness.

Namon (Thai) Lustral water.

Ngan sop look (Thai) "Counterfeit funeral".

Nibbana (Pali), ***niphan*** (Thai) **nirvana** (Sanskrit) Permanent extinction of suffering; release from the cycle of rebirth.

Pali Canonical language of Theravada Buddhism.

Paticcasamuppada (Pali) Doctrine of "dependent originations", explaining the causal relations between ignorance and craving and the arising of suffering and rebirth.

Phansa (Thai) Annual three-month rainy season retreat.

Phi (Thai), ***preta*** (Sanskrit), ***pret*** (Thai) Ghosts or spirits, some very tall and hungry.

Rot nam (Thai) Dedicate religious merit while pouring water.

Sai bat (Thai) Donate food to monks' and nuns' alms bowls.

Samsara (Pali) Phenomenal existence, the cycle of rebirth.

Sangha than (Thai) Offering consisting of food and money and, ideally, the laypersons that perform s*angha than* wear white.

Sati (Thai) Mindfulness or self-watchfulness.

Sikkhamat (Thai) A female ordained ascetic in the Asoke sect.

Stupa (Sanskrit) A reliquary monument.

Sala (Thai) Hall, pavilion, wayside shelter, study and merit-making hall, public building, preaching hall, multi-purpose hall.

Sai sin (Thai) Sacred cotton thread.

Samnak chi (Thai) *Mae chi*s' department at a temple, *mae chi*s' self-governed religious "abode".

Samsara (Pali) Phenomenal existence, the cycle of rebirth.

Sangha (Pali) The community of Buddhists. *Sangha* can be used just to refer to male monks, or to male and female monks, or it can also include laymen and women who have taken the five Buddhist precepts. In Thailand *sangha* refers to the assembly of male monks (*bhikkhu*).

Sawan (Thai) Heaven.

Tak bat (Thai) Donate food to monks' and nuns' alms bowls.

Thammada (Thai) To be normal, common, usual – referring to the reality of things.

Tham bun (Thai) To perform meritorious deeds.

Thewada (Thai) Devine angels.

Thipak song or ***samnak song*** (Thai) An unofficial/unsanctified "temple" or monastic centre lacking an ordination hall. *Samnak/thipak song*s commonly consist of a few *kuti*s (monks' and nuns' living quarters; often a small basic one-room house) and a *sala* (a multi-purpose hall).

Wai phra (Thai) Gesture of showing respect to the Buddha.

Wan phra (Thai) Buddhist day of worship.

Vassa (Pali) Annual three-month rainy season retreat; usually between July and October.

Wat (Thai) Buddhist temple

Winjan (Thai) Spirit, "soul".

Index

Page numbers in *italics* denote tables, those in **bold** denote figures.